The Four-Color Problem

ASSAULTS AND CONQUEST

THOMAS L. SAATY
University of Pittsburgh

and

PAUL C. KAINEN
McLean, Virginia

Dover Publications, Inc., New York

Published in Canada by General Publishing Company, Ltd., 30 Lesmill
Road, Don Mills, Toronto, Ontario.
Published in the United Kingdom by Constable and Company, Ltd.

This Dover edition, first published in 1986, is an unabridged, corrected
republication of the work originally published by the McGraw-Hill International Book Company, New York, 1977. In addition, the authors have provided a new Preface and a Supplementary Bibliography for this edition.

Manufactured in the United States of America
Dover Publications, Inc., 31 East 2nd Street, Mineola, N.Y. 11501

Library of Congress Cataloging-in-Publication Data
Saaty, Thomas L.
  The four-color problem.

  Bibliography: p.
  Includes index.
  1. Four-color problem.   I. Kainen, Paul C.   II. Title.   III. Title: 4 color
problem.
QA612.19.S2   1986   511'.5   85-29302
ISBN 0-486-65092-8

CONTENTS

PREFACE TO THE DOVER EDITION

In the years since the original publication of the Appel–Haken theorem, some things have happened that one might have expected. The experts are continuing to chip away at the length of time needed for the computation. Of course, computers are faster now. There has also been philosophical discussion about the desirability or purity of computer-assisted proofs. Methods are proposed to settle or at least check results without computer intervention.

Yet we must be somewhat perplexed at an overall *lack* of sustained reaction. The accolades one might have thought would be bestowed on the solvers of a famous mathematical enigma were never given. The elaborate conceptual and computational structure that Koch, Haken, and Appel built has been largely ignored. We believe that mathematicians get more excitement by doing the entire job of proving the existence of solutions and then constructing them by themselves than they do by relying on the computer as a substantial partner in the undertaking. Still, ambitious experts in artificial intelligence are telling us that it will not be long before computers will be doing all that we do, better and faster.

We hope this book will continue to evoke interest in the four-color problem, in its computer-aided solution, and perhaps in finding an alternative way to prove it. By the way, a natural follow-up would be a four-color *algorithm*.

For this edition we have corrected a number of errors and made a few minor revisions in the original text. We have also added a few items at the end of the Bibliography (see page 211).

Thomas L. Saaty
University of Pittsburgh

Paul C. Kainen
McLean, Virginia

PREFACE TO THE FIRST EDITION

A funny thing happened to this book on the way to the publisher. More than two years ago, we had completed the manuscript of a book entitled "Assaults on the Four-Color Conjecture." This book was based on the award-winning paper, "Thirteen Colorful Variations on Guthrie's Four-Color Conjecture," by one of us (T.L.S), which appeared in the *American Mathematical Monthly* in January 1972. Our original book was oriented towards the variety of approaches with which the four-color conjecture (4CC for brevity) has been investigated. The reader can well imagine our mixed reactions to the announcement, in the summer of 1976, that the 4CC had been proved—by a computer, no less.

This book represents our response to the challenge. We expanded the relevant sections and included a complete account of the innovative methodology which Kenneth Appel and Wolfgang Haken invented to solve the problem. This constitutes Part One of our present book. We have also retained a fairly thorough treatment of the impressive variety of approaches which mathematicians have used since the 1870s to try to solve the problem. This is Part Two of the book.

We do not plan to study the conjecture as a dry artifact. Rather, we see the myriad attacks and the method of solution as a vital force behind several important areas of modern mathematics.

The power and fecundity of thought brought to bear on the four-color problem is evidence, at least to us, that the presence of unsolved problems is a good thing. They bring out new ideas and stimulate novel approaches which otherwise might not emerge, and often these new methods become standard for attacking other problems.

Our observation with regard to the 4CC is that mathematicians can be divided into three groups: (1) those who were obsessed with the problem and concentrated their lives and energies on it; (2) those who maintained a balanced interest in it, every once in a while gathering their resources for a new attack; and (3) those who were passionately disinterested, as if they were afraid that facing the problem would necessitate a frustrated retreat. Several mathematicians of this century have belonged to the second group. They and others became deeply involved in trying to answer the 4CC, contributing significant variations and approaches which are interesting in their own right and which greatly enriched the whole of mathematics.

The four-color conjecture was one of the major unsolved problems in mathematics and has been attacked from a great diversity of standpoints, providing a one-of-a-kind pattern for meeting other challenging problems. It was like a forbidding fort being attacked and raided in numerous imaginative ways—ranging from tunneling through the earth, to mounting ladders, to using catapults, modern tanks, flame throwers, and even airplanes. The fort has finally collapsed, and even then it took a several-year, mounted attack using modern technology to do the job.

The sophisticated technique of Haken and Appel appears to have succeeded in proving that the 4CC is true. We say "appears to have succeeded" since their proof involves the computer-facilitated analysis of 1936 special cases, and will thus require several years for thorough checking. Even then, there will probably persist some lingering doubt among many scientists because of the elaborateness of the argument. Yet, with the success of their exhaustive method, they have, as we discuss in the text, established a new position for the computer in the realm of mathematical proof. They have also demonstrated to the mathematician that there is available a powerful tool which can supplement his attack on problems, not by merely constructing examples or counter-examples but by permitting him a much richer combinatorial repertoir including, for example, an entire catalog of solutions for all the special cases of his problem. This indeed involves a remarkable departure from traditional thinking in terms of simple and elegant proofs. Thus, even in solution, the four-color conjecture remains an enigma.

We would like to gratefully acknowledge the people who have helped us. Frank Bernhart has made a number of useful suggestions and has contributed generously of his time. We would also like to thank W. T. Tutte, John A. Koch, and Frank Harary for carefully reading the original manuscript. We are also indebted to Mary Lou Brown for her meticulous care in typing the manuscript.

Thomas L. Saaty
University of Pennsylvania

Paul C. Kainen
Case Western Reserve University

NOTE ON REFERENCES

The literature relevant to our subject has been divided into two parts. The first, containing books on graph theory, will be found on pages 194–196. The second part contains references on the four-color problem—a comprehensive collection of literature—and the reader is advised that references in the text to books should first be looked for in the first section.

T.L.S.
P.C.K.

THEME AND SOLUTION

HISTORICAL SETTING;
GUTHRIE'S CONJECTURE AND ITS DUAL

1-1 INTRODUCTION

The four-color conjecture (4CC) has been one of the great unsolved problems of mathematics. From 1852 and continuing to this day, practically every mathematician who has ever lived has, at one time or another, tried his or her hand at settling the conjecture. Many of the ideas which first developed to attack the 4CC have had important connections to other disciplines; for example, the entire field of graph theory has grown out of this fertile ground. Thus, the 4CC has had a seminal role in modern applied mathematics.

One of the transforming features of modern mathematics is the use of the digital computer. It is ironic but fitting that the 4CC now appears to have been proved by Wolfgang Haken and Kenneth Appel at the University of Illinois who used a well-orchestrated approach that involved, among other things, 10^{10} separate operations on a high-speed computer to prove a finitistic form of the conjecture.

However, the issue remains clouded because of the staggering length of the calculations (1200 hours in total) and because of the elaborate argument needed to understand how the machine computations solve the problem. Such a lengthy program (both computer instructions and mathematical reasoning) certainly requires careful verification which may be years away.

Mathematical interest in this proof has been intense, both because of the fame of the problem and the novel methods employed. Much of the attention is frankly skeptical. A sequence of minor programming errors has been found and while, so far, each has been eliminated with trivial effort, there remains the possibility that one will turn up which necessitates a major revision in the proof or invalidates it altogether.

But the Appel–Haken *theory* argues on the basis of several heuristic assumptions that there is an overwhelmingly great probability that their method of proof must succeed. Since the proof itself (if it is a proof) was discovered using the theory, and since the theory is heuristic, there is an added tendency on the part of many mathematicians to mistrust the whole thing. On the other hand, those who believe the theory are inclined to believe the proof. This controversy may not be settled soon. Our purpose here is to present the mathematical history of the four-color conjecture including the proposed proof and its underlying heuristic theory, as well as the earlier ideas of Heesch and Birkoff which form the basis of most modern research on the problem. We have also included the interesting and varied approaches which have been taken to equivalent forms of the 4CC. The ultimate decision on the Appel–Haken proof will be made by the mathematical and scientific community. The history of the 4CC amply demonstrates how often we have been wrong in the past.

> The *four-color conjecture* says that four colors are sufficient to color any map drawn in the plane or on a sphere so that no two regions with a common boundary line are colored with the same color.

Two regions which have a point or a finite number of points in common are permitted to have the same color. It is also required that all regions be connected together.

Virtually every important variant of the four-color problem is completely solved—even problems involving more than four colors. For example, any map on the torus (surface of a doughnut) can be colored with at most seven colors and there are toroidal maps which require seven colors.

That four colors are necessary can be seen from the two maps below (see Fig. 1-1). The first has four mutually adjacent regions and, hence, obviously requires four colors. However, this type of condition need not hold in order that four colors be necessary, as illustrated by the second map.

The first map illustrates a *local obstruction* to three-coloring a map; i.e., there is some set of four regions which are pairwise mutually adjacent. On the other hand, the second map is an example of a *global obstruction* to three-coloring a map. If the central country is removed and the remainder of the map is three-colored, then all

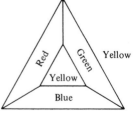
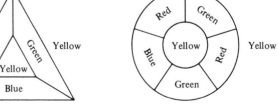

Figure 1-1

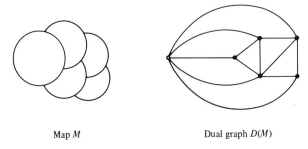

Map *M*   Dual graph *D*(*M*)

Figure 1-2

three colors would be used for neighbors of the central region which would, hence, require a fourth color.

It is interesting to note that in trying to four-color a map, we shall find that there *is* no local obstruction; one cannot have five mutually adjacent regions. With the solution of the problem, we also know that no global obstructions can exist.

It turns out to be quite useful to consider a different kind of coloring problem intimately related to that of coloring a map. Place a point, or *vertex*, in the middle of each country of some map *M* and join two vertices with a line, or *edge*, whenever the two countries have a common border, as indicated above in Fig. 1-2, to obtain a *dual graph D*(*M*). Call two vertices of *D*(*M*) *adjacent* if they are joined by an edge. The vertices are called the *endpoints* of the edge. An edge is said to be *incident* to its endpoints and vice versa.

The problem of four-coloring the regions of a map may now be stated in terms of coloring the vertices of its dual graph so that, whenever two vertices are joined by a line, they are colored differently. Thus, the 4CC as stated in its regional form is equivalent to the statement that we can four-color the vertices of certain kinds of graphs; namely, those which are dual to maps. In fact, we will see later that any connected graph which can be drawn in the plane is dual to some map. It is this latter formulation of the 4CC which allows us to view it as a problem in graph theory.

By its very construction, any dual graph *D*(*M*) has the property of being *planar*; i.e., we can represent its vertices and edges in the plane so that edges cross one another only at common endpoints. Since graphs are more general than maps, we can consider the problem of coloring any graph in some special class. For example, it turns out that coloring planar graphs is exactly as difficult as coloring maps. We can also consider the problem of coloring special types of maps, as well as some problems which arise by varying our notion of a map to permit maps on other surfaces than the plane (or sphere) or to allow disconnected countries (empires).

Thus, for any class of graphs or maps, we can ask what is the smallest possible number of colors needed to color any graph or map in the collection. This is an existence question: "What is the smallest number, if one exists, which suffices to color any graph or map subject to some constraint?" Answering this question does not prepare us to answer a second, very practical question: "How does one actually find a coloring of some particular graph or map?" More simply: "Given a map, color it with as few colors as possible."

We prove that every planar map can be colored with at most five colors. The map in Fig. 1-3 can, in fact, be four-colored, but it is not at all obvious how to find such a coloring and the reader may start the warm-up by trying to color it (using the numbers 1, 2, 3, 4 instead of colors to preserve the book).

Brief Historical Account

After careful analysis of information regarding the origins of the four-color conjecture, Kenneth O. May (1965) concludes that: "it was not the culmination of a series of individual efforts, but flashed across the mind of Francis Guthrie while coloring a map of England ... his brother communicated the conjecture, but not the attempted proof to DeMorgan in October, 1852." His information also reveals that DeMorgan gave it some thought and communicated it to his students and to other mathematicians, giving credit to Guthrie. In 1878 the first printed reference to the conjecture, by Cayley, appeared in the *Proceedings of the London Mathematical Society*. He wrote asking whether the conjecture had been proved. This launched its colorful career involving a number of equivalent variations, conjectures, and false proofs. But the question of sufficiency, which had been wide open for such a long time, is now closed.

One of the many surprising aspects of the 4CC is that a number of the most important contributions to the subject were originally made with the belief that they were solutions. After Cayley publicized the problem in 1878, Kempe (1879b) published a "proof" of the conjecture. Eleven years later, Heawood (1890) pointed out the error in Kempe's argument and salvaged enough to prove the sufficiency of five colors. In the meantime, Story (1879) used Kempe's work to reduce the conjecture.

In 1880, Tait also "proved" the conjecture. His proof, while based on the false

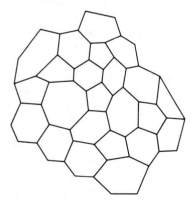

Figure 1-3

assumption that every three-connected planar graph is Hamiltonian, introduced the important notion of three-coloring the edges of a cubic map. Petersen (1891) realized that Tait's methods really showed that the 4CC was equivalent to a conjecture on coloring edges. Refining Tait's argument, Petersen also showed that the 4CC is equivalent to the conjecture that any planar cubic map can be toured by a Hamiltonian circuit or by a collection of mutually disjoint subcircuits of even length. Heawood (1898) transformed the conjecture to algebraic form, while Veblen (1912) transformed the conjecture into equivalent assertions in projective geometry. In the same year, Birkhoff (1912) introduced the very important notion of chromatic polynomials which has also been studied by Birkhoff and Lewis, Whitney, and, more recently, Tutte.

The notion of dual graph mentioned above was introduced by Whitney (1931) and used to give an elegant characterization of when a graph is planar. In particular, his arguments included the basic ideas of matroid theory. Whitney also proved that every maximal planar graph without separating triangles has a Hamiltonian circuit, and used this to reduce the 4CC for arbitrary planar graphs to Hamiltonian planar graphs.

In 1922, Franklin showed that every map with 25 or fewer regions can be four-colored. The basic technique in his and in subsequent work in this area is to show that certain configurations are reducible; i.e., can never occur in irreducible maps. In fact, probably the first important reducible configuration is due to Birkhoff a few years earlier. Franklin's (1922) work showed that any proposed counter-example to the 4CC would need at least 26 regions.

In 1925, Errera noted that any map satisfying Franklin's lower bound of 26 regions would have at least 13 pentagons. In 1926, Reynolds raised the lower bound to 28. In 1937, considering a paper of Franklin to be published in the following year (1938) in which Franklin raised the lower bound to 32, Winn showed that there must be at least two regions with more than six sides in such a map. In his paper, Franklin (1938) also showed that any irreducible map satisfying the lower bound of 32 needed at least 15 pentagons. Winn (1940) raised the lower bound, once more, to 36, where it remained for 20 years until Ore and Stemple (Ore, 1970) raised it to 40.

Other important contributions include Brooks' (1941) theorem, in which he gave a bound on the chromatic number of any graph, and Hadwiger's (1943) conjecture, of which the 4CC is a special case. We should also refer to the works of Dirac on critical graphs and of Tutte in a variety of areas.

As interest in the 4CC has increased, so has interest in graph theory. König published the first book on graph theory in 1936. There are now several other general texts by Berge (1961 and 1973), Busacker and Saaty (1965), Harary (1969), Ore (1961 and 1963) and Roy (1969), not to mention Tutte's specialized mono-graph (1966) on connectivity and numerous symposia and colloquia proceedings.

In 1952, Dynkin and Uspenski published a small book of elementary coloring exercises. Ringel published a major book on coloring graphs and maps seven years later, in 1959, and another one in 1974, and in 1967, Ore's now classic book on the subject appeared. In 1969, Heesch wrote an important monograph on one aspect of the problem; it dealt with the subject of charging and discharging and reducible configurations. In the meantime, the number of regions in an arbitrarily drawn map that can be four-colored was extended to 52 and, finally, to 96. But then the current solution of the problem came in 1976.

One particularly significant thing which has emerged from this approach is that the computer will play an increasingly fundamental role in mathematical thinking. Here it has been an invaluable aid in generating exhaustive analyses of a problem by cases. From now on mathematicians will dare to use the computer more and more in constructing proofs. As this use becomes more prevalent it would gain greater credibility as a method of approach. This says that we may more often look for constructive analytical existence proofs by working out cases instead of nonconstructive synthetic ones. With this, one would expect that vastly larger and more detailed problems of an abstract nature significant to man's logical view of his thought processes and to his role in the universe will be formulated and solved. Man's imagination creates the domain within which the computer will function.

As we have seen, the 4CC is easy to state. The challenge it usually evokes, if pursued enthusiastically, provides a quick and rather intensive entry to graph theory. Although enthusiasm is helpful in solving problems, it does not guarantee results. The great mathematician, Herman Minkowski, once told his students that the 4CC had not been settled because only third-rate mathematicians had concerned themselves with it. "I believe I can prove it," he declared. After a long period, he admitted, "Heaven is angered by my arrogance; my proof is also defective."

It is our hope to evoke such interest and enthusiasm for understanding this problem. We believe that our exposition is sufficiently elementary to make the ideas accessible to mature beginners. After we finish with the rather direct route of showing the essential lines that the proof followed in the general scheme of things, we give a relatively thorough presentation of equivalent formulations. We also deal with important related areas such as chromatic numbers and chromatic polynomials that we believe will be here to stay for other good uses. However, we have deliberately refrained from exploring *every* subject related to the 4CC for this could include the entire field of graph theory.

The logical organization of the book is summarized in the chart of Fig. 1-4. The dark arrows show the route which the solution followed between 1879 and 1976. The double arrows indicate equivalence between the abbreviated statements. Maps are on one side of the page and their graphs on the other, with a duality line separating them in the middle. Note that the solution came by following what we call the canonical way—reducing maps to triangular networks in the dual setting and solving the problem for such networks. The diagram shows pockets of activity in various areas which seemed promising to the researchers. For example, from cubic maps we have edge-coloring formulations. The study of coloring maps in surfaces had the hope of producing the result. From the central circle representing the map-coloring problem and its dual sprouted many equivalent formulations, each hoping to shed greater light on the problem from a completely new

9

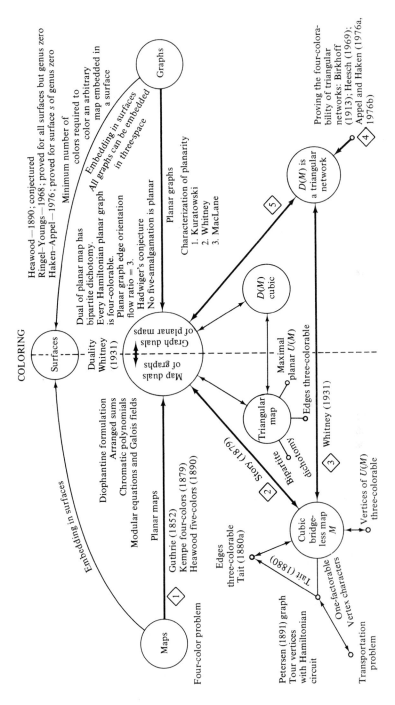

Figure 1-4

vantage point. Solving the problem meant breaking open one of these equivalent forms and showing that the 4CC was true for it. This appears to have happened in one place shown in the chart. As a consequence, all the others will be also true (by equivalence).

In the remainder of this chapter, our plan is as follows: Sec. 1-2 gives basic definitions and concepts from the theory of maps and graphs. We formulate the 4CC as a conjecture about planar graphs, as well as maps, and prove the equivalence of these notions.

One of the most important tools in combinatorial topology and the center of all topological approaches to coloring is Euler's formula. We present in Sec. 1-3 an elementary derivation, via induction and some standard properties of trees. Planarity is also discussed, as well as its implications for a relationship between the number of edges and vertices in a planar graph. Instead of proceeding to Chap. 2, the reader anxious to learn about the Appel–Haken proof may go directly to Chap. 3, referring to Chap. 2 when necessary for understanding.

In Chap. 2 we systematically examine the general problem of coloring in a graph or a map. Some of the major topics which we cover include Brooks' theorem, the five-color theorem, and the work of Ringel and Youngs which settled the analogous problem to the 4CC on all surfaces but the plane (or sphere).

Chapter 3 represents the focus of the book. It is here that we develop the basic theory and the relevant detail necessary to understand the proposed proof of the four-color theorem by Appel and Haken. In addition to treating discharging and reducibility, we show how and why the computer is used in these arguments. In closing, the chapter delves into the new probabilistic and heuristic methodology which led to the proof's discovery.

In Part Two, the last four chapters of the book give a large number of equivalent formulations developed over the past century. These include such diverse areas as algebra, number theory, and finite geometries. Many of these alternate forms have produced valuable mathematics. We are permitted now, with hindsight and a solution of the problem, to see the overall plan.

1-2 DEFINITIONS—GRAPHS, MAPS, AND THE 4CC

We now define many of the basic concepts used throughout the book including graphs, maps, and dual graphs. We also prove that the 4CC for maps is equivalent to the 4CC for graphs. Essentially, a graph is a set of points and line segments, or simple curves of which they are the endpoints. But we need more precision, as we also need to know which lines join which points.

If S is a set, we write $|S|$ for the cardinality of S. In particular, $|S| < \infty$ means S is finite and then $|S|$ is simply the number of elements in S. If S and T are sets, $S \times T$ denotes the cartesian product of S and T; that is, the set of all ordered pairs (s, t) with $s \in S$ and $t \in T$. As usual, $(s, t) = (s', t')$ if and only if $s = s'$ and

$t = t'$. If S is a set, we define $S \& S$, the *unordered product* of S with itself, to be the set of all unordered pairs $[s, s']$, $s, s' \in S$. Each unordered pair $[s, s']$ is the equivalence class of an ordered pair (s, s') under the relation $[s, s'] = [s', s]$. The only difference between these unordered pairs and the sets with two elements is that $[s, s]$ is an unordered pair while $\{s, s\} = \{s\}$ has only one element.

A *graph* G is a triple (V, E, Φ), where $|V|$ with $0 < |V| < \infty$ is called the *order* of the graph, $|E| < \infty$, and $\Phi : E \to V \& V$ is a function. We call $v \in V$ a *vertex* or *point*, $e \in E$ an *edge* or *line*, and Φ the *incidence function*. If $\Phi(e) = [v, w]$, we say that e is *incident* to v and to w and that v and w are *adjacent*; v and w are also called the *endpoints* of e. We sometimes write $V(G)$ or $E(G)$ for the set of vertices or edges of graph G. If edges e and e' have a common endpoint, they are called *adjacent*; if they have two common endpoints, they are *parallel edges*. An edge with only one endpoint is called a *loop*. A graph is *simple* if it has no loops or parallel edges.

Note that if $G = (V, E, \Phi)$ is a simple graph, then Φ is a one-to-one function and so we may identify E with a subset of $V \& V$. Thus, G is determined by the pair (V, E) alone and we write $G = (V, E)$.

Let $G = (V, E, \Phi)$ be a graph and consider subsets $W \subset V$ and $F \subset E$ where $W \neq \theta$. Obviously, $W \& W$ is a subset of $V \& V$. If $\Phi(F) \subset W \& W$, then the triple $(W, F, \Phi | F)$ is a graph which we call a *subgraph* of G, where $\Phi | F$ is the function from F to $W \& W$ induced by Φ. Thus, a subgraph $G' = (V', E', \Phi)$ of G is determined by a (nonempty) subset V' of V and a subset E' of E, provided that if v is any vertex in V which is an endpoint of some edge e in E' then $v \in V'$. The function Φ' is just $\Phi | E$. If $V' = V$, the subgraph is called *spanning*. A subgraph is called a *section (induced, full)* subgraph if any two vertices in the subgraph are joined by as many edges (in the subgraph) as they are in the larger graph.

For example, in Fig. 1-5 below, H_1 is a subgraph of G but not a full subgraph. On the other hand, H_2 is full and H_3 is spanning. Note that a full subgraph G' of G is completely determined by its set of vertices.

The following definitions are concerned with a tour of edges of the graph. There are several possibilities and we need to distinguish among them.

A sequence of n edges $e_1, \ldots, e_n$ in a graph G is called a *walk* or *edge progression* if there exists an appropriate sequence of $n + 1$ (not necessarily distinct) vertices $v_0, v_1, \ldots, v_n$ such that e_i is incident to v_{i-1} and to v_i, $i = 1, \ldots, n$. A walk is *closed* if $v_0 = v_n$ and *open* otherwise. If $e_i \neq e_j$ for all i and j, $i \neq j$, the walk is called a *trail* or *chain*. A closed chain is called a *circuit*. A set of edges is also referred to as a *trail*.

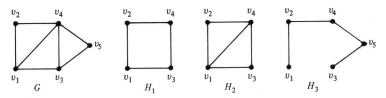

Figure 1-5

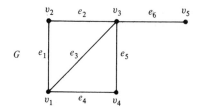

Figure 1-6

If all of its vertices are distinct, a walk is called a *simple chain* while, if $v_0 = v_n$ and all other vertices are distinct, we have a *simple circuit* provided $n \geq 3$.

The *degree* (or *valence*) of a vertex v is an important numerical way to distinguish between the vertices. It is the number of edges incident with that vertex and is denoted by $\deg(v)$. A loop counts twice. We write $\delta(G)$ for $\min\limits_{v \in V(G)} \deg(v)$ and $\Delta(G)$ for $\max\limits_{v \in V(G)} \deg(v)$.

It is interesting to note that there is an even number of vertices in any graph, whose degree is odd. To see this, let m be the number of edges and n the number of vertices; then $2m = \Sigma \deg(v) = \Sigma \text{ odd } \deg(v) + \Sigma \text{ even } \deg(v)$. We use $2m$ because each edge is counted twice. It follows that Σ odd $\deg(v)$ must have an even number of terms.

In Fig. 1-6, $\deg(v_3) = 4 = \Delta(G)$, while $\deg(v_5) = 1 = \delta(G)$. The sequence e_1, e_2, e_3, e_1, e_2 is a walk from v_1 to v_3, while e_1, e_2 is a chain. The sequence e_1, e_2, e_5, e_4 is a circuit.

Central to our study of the 4CC is to recognize a graph which can be drawn in the plane. It is easy to find a graph, some of whose edges always cross at a point that is not a vertex no matter how we draw them. We need ways which tell us when a graph is planar. Such characterizations will be given later.

A graph is *planar* if it can be embedded (drawn) in a plane such that no two edges meet except at a common endpoint. Again a graph may be in one or in several pieces called *components*. It is in one piece or *connected* if every pair of vertices can be joined by a walk. A *connected component* of a graph is a maximal connected subgraph. A graph is *complete* if it is simple and every pair of distinct vertices are adjacent (a complete graph with n vertices is denoted by K_n); *k-partite* if its vertices can be partitioned into k disjoint sets so that no two vertices within the same set are adjacent; and *complete k-partite* if every pair of vertices in different sets are adjacent. If $k = 2$, we call the graph *bipartite*, and if the graph is complete bipartite, we denote it by $K_{p,q}$, where p and q are the number of vertices in the two disjoint vertex sets. A complete three-partite graph is shown in Fig. 1-7.

A *map* or *planar map* M consists of a connected planar graph G together with a particular drawing or embedding of G in the plane. We call G the *underlying graph* of M and write $G = U(M)$. The map M divides the plane into connected components which we call the *regions*, or *faces*, or *countries* of the map. Two regions are *adjacent* if their boundaries have at least one common edge, not merely a common vertex. We refer to the edges in the boundary of a region as its *sides*.

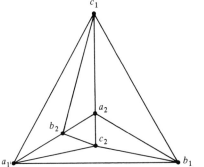

Figure 1-7

Note that one graph may be embedded in the plane to produce several different maps; i.e., different maps can have the same underlying graph. For example, in Fig. 1-8, the graph which consists of a square and two triangles all meeting at one vertex is embedded in the plane in two ways: one has both triangles inside the square; another has one inside and one outside the square. In the second map there is no four-sided region, while in the first map, the region exterior to the square has four sides. Later on, however, we will see that this phenomenon cannot occur in interesting cases.

A *k-coloring* of a map (graph) (sometimes called a *proper k-coloring*) is an assignment of k colors to the regions of the map (vertices of the graph) in such a way that no two adjacent regions (vertices) receive the same color. A map (graph) is *k-colorable* if it has a k-coloring. One can also define an *edge-coloring* of a graph (or map) by assigning colors to the edges so that adjacent edges (or edges bounding different regions) receive different colors. Later, various relationships between coloring vertices, edges, and regions will be developed. See Fig. 1-9 for an example of a map whose regions are four-colored, whose edges are three-colored, and whose vertices are three-colored.

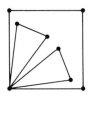
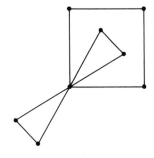

Figure 1-8

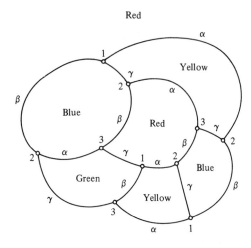

Figure 1-9

Suppose that $G = (V, E, \Phi)$ and $G' = (V', E', \Phi')$ are graphs. A homomorphism $\Psi: G \to G'$ is a pair of functions $\rho: V \to V'$ and $\sigma: E \to E'$ such that if $[v, w] = \Phi(e)$, then $[\rho v, \rho w] = \Phi'(\sigma(e))$. If G and G' are simple, then Ψ is completely determined by ρ. Thus, a function $\rho: V \to V'$ determines a homomorphism if ρv and ρw are adjacent in G' whenever v and w are adjacent in G. In particular, if v and w are adjacent, then $\rho v \neq \rho w$. A homomorphism $\Psi = (\rho, \sigma)$, $\Psi: G \to G'$ is an *isomorphism* if ρ and σ are both one to one and onto. We also say that G and G' are *isomorphic*. Essentially, an ismorphism is just a relabeling of the vertices and edges. Let $G = (V, E, \Phi)$ be a graph. An *elementary subdivision* of G is obtained by replacing some edge e, whose endpoints are v and w, with edges e' and e'', where e' has endpoints v and x_e and e'' has endpoints x_e and w. Thus, we add one new vertex x_e and replace one old edge e with two new edges e' and e''. G' is a *subdivision* of G if there is a finite sequence of graphs $G_1, \ldots, G_k$ where G_{i+1} is an elementary subdivision of G_i $(1 \leq i \leq k - 1)$ and $G_1 = G$, $G_k = G$. Two graphs G and H are *homeomorphic* if there exist subdivisions G' of G and H' of H such that G' and H' are isomorphic.

Roughly speaking, two graphs are homeomorphic when they are isomorphic up to vertices of degree 2. For example, in Fig. 1-10, G_1 and G_2 are homeomorphic.

Two maps are *isomorphic* if there is a continuous deformation of the plane onto itself carrying the vertices, edges, and regions of one map onto the vertices, edges, and regions of the other. In particular, infinite regions correspond. Isomorphic maps have isomorphic underlying graphs.

The *dual graph* $D(M)$ to a map M is constructed as follows. The set of vertices of $D(M)$ is in one-to-one correspondence with the set of regions of M. Let us denote the correspondence from regions to vertices by $R \leftrightarrow \hat{R}$. If R and S are (not necessarily distinct) regions of M, then we join $\hat{R}$ and $\hat{S}$ by one edge $\hat{e}$ for every edge e which lies on the common boundary of R and S. The edge $\hat{e}$ is

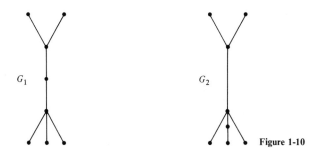

G_1 G_2 Figure 1-10

called *orthogonal* to e. The most important fact about dual graphs is that they are planar.

Theorem 1-1 $D(M)$ is planar.

PROOF We simply show how to draw $D(M)$ in the plane. Choose one point $\hat{R}$ in the interior of each region R. If two regions R and S have a common boundary edge e, then connect the points $\hat{R}$ and $\hat{S}$ by a simple arc $\hat{e}$, with endpoints $\hat{R}$ and $\hat{S}$, which passes through the interior of the edge e. We illustrate this in Fig. 1-11. Although there is some choice as to where to choose the points $\hat{R}$ and the curves $\hat{e}$, the resulting map M^* is unique (up to isomorphism) and we call it the *dual map* to M. Thus, $D(M) = U(M^*)$.

We are now in a position to state the four-color conjecture (4CC).

Conjecture C_0(Guthrie) Every map can be four-colored.

We want to prove now that this conjecture is equivalent to another conjecture about coloring planar graphs. Using the notion of dual graph, the next statement converts a map-coloring problem into a graph-coloring problem.

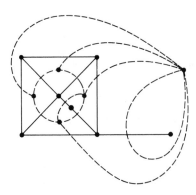

Figure 1-11

Conjecture C_1 Every planar graph is four-colorable.

Theorem 1-2 Conjectures C_0 and C_1 are equivalent.

PROOF Let M be any map. Then $D(M)$ is a planar graph; so by C_1 the vertices of $D(M)$ can be four-colored. But now, using the one-to-one adjacency-preserving correspondence between vertices of $D(M)$ and regions of M, we can simply transfer the coloring of $D(M)$ to a coloring of M. This shows that C_1 implies C_0.

Now we must prove that C_0 implies C_1. So let us assume that conjecture C_0 holds. It will clearly suffice to show that every connected planar graph can be four-colored. The usual approach, due to Whitney (1933), is to show that every connected planar graph G is isomorphic to $D(M)$ for some map M. This requires the result that $(N^*)^*$ is isomorphic to N for any map N. (We will prove this result later.) For the time being, we prefer to give a different proof.

Construction Let M be any map. Draw a small circle C_v around every vertex v of M and erase the interior of C_v including v. We may assume without loss of generality that C_v contains no vertices other than v, that C_v intersects only those edges of M which have v for one of the endpoints, and that if C_v intersects an edge e, then C_v and e intersect in a unique point which we call e_v. Define a new map $T(M)$ as follows. The regions of $T(M)$ consist of the regions of M together with the regions R_v interior to the circles C_v. (By making the circles C_v sufficiently small, we may insure that the regions R_v do not intersect each other.) The vertices of $T(M)$ consist of the points e_v, and the edges of $T(M)$ consist of the segments $\bar{e}$ of edges e contained between the vertices e_v and e_w where $[v,w] = e$ and the segment (e_v,f_v) contained in C_v between the vertices e_v and f_v whenever e "follows" f at v. We say that $T(M)$ is obtained from M by "inflating" vertices (see Fig. 1-12). We need a further modification $\bar{M}$ of $T(M)$ which is indicated below. One may think of $\bar{M}$ as being obtained from M by "thickening" the edges of M.

The key property of $\bar{M}$ for our purposes is that it contains a region $\bar{v}$ for every vertex v in M and that regions $\bar{v}$ and $\bar{w}$ are adjacent in $\bar{M}$ if and only if the vertices v and w are adjacent in M. Thus, $U(M)$ is a full subgraph of $D(\bar{M})$.

Now we are ready to finish the proof that C_0 implies C_1. Let G be any planar graph. Drawing G in the plane produces a map M whose underlying graph is G. According to C_0 we can four-color the thickened map $\bar{M}$. Now coloring each vertex v of G with the color used for the region $\bar{v}$ of $\bar{M}$ provides us with a four-coloring of G.

1-3 EULER'S FORMULA (A NECESSARY CONDITION FOR PLANARITY)

In all of mathematics, there are few formulas as simultaneously elegant and useful as Euler's formula. This formula relates the numbers n, m, and r of vertices,

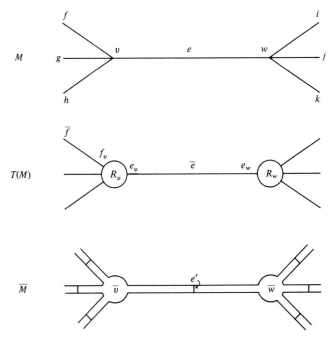

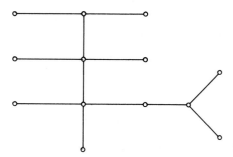

Figure 1-12

edges, and regions, respectively, in a planar map and states that $n - m + r = 2$. In an appendix we will give several necessary and sufficient theorems for planarity. Euler's condition is easy to derive and happens to serve our needs.

Before proving Euler's formula, however, we shall need to develop two other concepts: trees and bridges. A *tree* is a connected graph with no circuits (see Fig. 1-13). Trees occur frequently in nature; e.g., many chemical compounds can be represented by trees. A tree edge is called a *branch* and a nontree edge is called a *chord*.

Figure 1-13

Theorem 1-3 Let G be a tree with n vertices and m edges. Then $n - m = 1$.

PROOF Use induction on m. If $m = 0$, the tree consists of a single vertex and the formula holds. Now suppose that m is any positive integer and assume that the formula holds for $m - 1$. Since m is positive, every vertex in the tree has a positive degree (otherwise G would not be connected). If $\delta(G) \geq 2$, then one can easily verify that G would contain a circuit. Therefore, there is at least one vertex of degree 1. Call the vertex v and let $H = G - v$.

Then H is a tree with $m - 1$ edges and $n - 1$ vertices and so, by induction, $(n - 1) - (m - 1) = 1$. But $n - m = (n - 1) - (m - 1)$.

The converse of this theorem is also true; i.e., if G is a connected graph with n vertices and m edges and $n - m = 1$, then G is a tree. We can always remove edges from a connected graph until we obtain a tree with the same number of vertices as the original graph and with m' edges. If G is not a tree, then $m' \neq m$ so $1 = n - m' \neq n - m$.

Trees with more than one vertex have another interesting property: the removal of any edge leaves a disconnected graph. This leads us to define a *bridge* as an edge in a connected graph whose removal disconnects the graph (see Fig. 1-14.)

The following lemmas are very easy to prove.

Lemma 1-1 Let e be an edge of a connected graph G. Then e is a bridge if and only if e lies on no circuit.

Lemma 1-2 Let M be a map and let e be an edge of G. Then e is a bridge if and only if e lies on the boundary of only one region.

Now we are ready for Euler's formula.

Theorem 1-4 Let M be a map whose underlying graph G has n vertices and m edges and suppose that M has r regions. Then $n - m + r = 2$.

PROOF Use induction on m. If $m = 0$, the formula holds since then $n = 1$ and $r = 1$. Assume the theorem is true for $m - 1$. We now try to prove it for m. There are two possibilities to consider: either M has an edge which is not a bridge or it does not. In the latter case, G is a tree. But then $r = 1$ since G has no circuits, and, by Theorem 1-3, we have $n - m = 1$ so the formula holds.

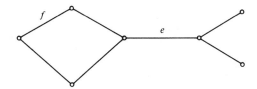

e is a bridge; f is not a bridge **Figure 1-14**

Suppose there is an edge e which is not a bridge. Let G' be the graph obtained from G by removing e, and let M' be the corresponding map. Then G' is connected and has $m - 1$ edges so, by induction, $n - (m - 1) + r' = 2$, where r' is the number of regions in M'. But since e is not a bridge, it must separate two distinct regions and thus $r' = r - 1$. Now $n - (m - 1) + r' = n - (m - 1) + (r - 1) = n - m + r$, which completes the proof.

Corollary 1-1 Let G be a simple connected planar graph with n vertices and m edges. Then $m \leq 3n - 6$.

PROOF Let M be a map with r regions such that $U(M) = G$. Since G is simple, every region of M has at least three sides. Moreover, every edge of M lies on the boundary of at most two regions. Therefore, $3r \leq 2m$. Hence, $2 = n - m + r \leq n - (m/3)$; that is, $m \leq 3n - 6$.

The same argument can be used to prove the following more general result. If G is not a tree, its *girth* is the length of its shortest circuit.

Corollary 1-2 Let G be a simple connected planar graph with n vertices and m edges which is not a tree. Then $m \leq [\gamma/(\gamma - 2)](n - 2)$, where γ is the girth of G.

Theorem 1-5 The complete graph on five vertices K_5 and the bipartite graph $K_{3,3}$ are not planar (see Fig. 1-15).

PROOF From Corollary 1-1 we have $10 \leq 9$ and, hence, K_5 is not planar. Similarly, Corollary 1-2 implies that, if $K_{3,3}$ were planar, then since the girth of $K_{3,3}$ is 4, we would have $9 \leq 2(6 - 2) = 8$, which is a contradiction. Alternately, if $K_{3,3}$ were planar, on substituting $n = 6$, $m = 9$ in Euler's formula, we would have $r = 5$. No region can have a triangle for the boundary for then there would be an edge joining two vertices in one of the two sets of the

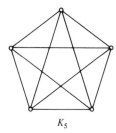

K_5

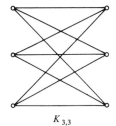

$K_{3,3}$

Figure 1-15

bipartite graph, contradicting its definition. Thus, every region is bounded by at least four edges, each of which is counted twice. Thus, $4r \leq 2m$ or $m \geq 2r$. But $m = 9$ and $r = 5$ and thus $9 \geq 10$, a contradiction. Thus, $K_{3,3}$ cannot be planar.

$K_{3,3}$ is known as the utility graph in which three houses are each connected to three utilities—gas, electricity, water. The object is to connect them so they do not intersect, an impossible task as shown above.

Both K_5 and $K_{3,3}$ play a central role in the characterization of planarity by Kuratowski (1930). This is discussed in the Appendix to this book (see page 176).

MAP COLORING;
PROBLEMS AND METHODS

2-1 INTRODUCTION

This chapter will serve the purpose of a systematic introduction to elementary coloring problems and their solutions. It is intended to provide a modicum of familiarity with the ideas of the subject and, in particular, with its methods.

The first type of question we examine (in Sec. 2-2) concerns the characterization of the structure of graphs and maps that are colorable with one, two, or three colors. But this subject does not generalize beyond such a small number of colors. In particular, there is no such characterization of four-colorability.

The second type of question we examine has to do with the minimum number of colors, called the chromatic number, that is sufficient for coloring a given graph. We give general bounds on this number for an arbitrary graph. Also, in Sec. 2-3, we show in a few lines the sufficiency of six colors for coloring any map or planar graph. The long-standing method of Kempe (1879b), that proved faulty for four colors but which was refined by Heawood (1890), is used to prove that five colors are sufficient for coloring an arbitrary planar graph. We then move to the torus and show that seven colors are sufficient and sometimes necessary for coloring graphs on that surface. Eight colors and more occur in conjunction with higher genus surfaces and empires.

In Sec. 2-4 we consider the third type of question. What is the minimum number of colors sufficient to color any graph or map embedded in a given surface? This number is called the Heawood number of that surface. It is in relation to a planar surface that the four-color problem arises. The work of Ringel and Youngs (1969) gives the answer for all surfaces but the plane. It turns out that an upper bound for the Heawood number associated with the surface is easily computed by an algebraic technique depending on Euler's formula for that surface. This number is equivalent to the largest possible chromatic number of a graph embedded in that surface. Although we discuss the Heawood number for surfaces, we do not give here the recent and lengthy proof by construction which shows that the upper bound can always be reached (see Ringel, 1974). This is just the reverse of our problem. It is easy to give examples of planar graphs requiring four colors. The upper bound part of the argument is the hard part in the plane, for, unfortunately, the upper bound argument on Heawood's number is not applicable to the plane.

The last section of the chapter gives an algorithm for coloring any map, not necessarily with the exact number of colors, but at least it gives a way for coloring which is, so far, one of the easy "good" ways for doing it.

2-2 ONE-, TWO-, AND THREE-COLORING

In this section we investigate one-, two-, and three-coloring of graphs and maps. Although these results are mostly much easier than what is to follow, they lead naturally to some of the basic ideas in graph theory and suggest why coloring is important.

What would it mean to be able to color a map M with only one color? A moment's reflection shows that M must be a *degenerate* map consisting of a single (unbounded) region, for if M contains a bounded region R, the edges in the boundary of R must separate R from some other region S, and R and S must receive different colors. For example, such a map might look like that shown in Fig. 2-1.

We could adopt "graph" as the basic notion and define a "map" to be a particular drawing of some (planar) graph. This approach has certain advantages. For example, one may be inclined to exclude maps, as in Fig. 2-2. One would be hard put to banish this kind of map without invoking nonsimple connectivity or some other slightly esoteric topological stigma. But, in the above parlance, we can merely insist that $U(M)$ be a *connected* graph. This means that one can join any two points (vertices) in $U(M)$ with a chain lying entirely within $U(M)$.

Figure 2-1

Figure 2-2

Suppose that $U(M)$ contains a circuit G. Since C is a simple closed curve it must divide the plane into two regions. Conversely, if M has two or more regions, then at least one of them must be enclosed by a circuit. The presence of circuits proliferates the number of adjacent regions. So we can conclude that M can be one-colored if and only if $U(M)$ contains no circuit.

But we have a name for a graph that is both connected and has no circuits. It is a tree. Thus, we have proved the elementary fact:

Theorem 2-1 The regions of a map M are one-colorable if and only if $U(M)$ is a tree.

The situation for graphs is even more simple (see Fig. 2-3). It is clear that if any two vertices have an edge in common, they would need two colors.

Theorem 2-2 The vertices of a graph G are one-colorable if and only if G contains no edges.

Of course, it follows that a connected graph is one-colorable if and only if it consists of a single vertex. In that case it is called *trivial*. This allows us to reestablish the fact that a one-colorable map M has a single region. To see this we note that if M is one-colorable, so is $D(M)$. But $D(M)$ is a connected graph and, therefore, it has a single vertex; but then M has a single region.

The situation for two-colorability is slightly more complicated but still manageable. When is a map two-colorable? Everyone's favorite example of a two-colorable map is the checkerboard (see Fig. 2-4), but this is really misleading. What should we color the external region since it touches both black and white squares?

Of course, the infinite checkerboard does not present this problem, but we are interested in maps with a finite number of regions. Thus, here is a better example of a two-colorable map. If we observe Fig. 2-5 carefully, we see that every point of the map meets an even number of lines. The number of lines which meet at a vertex is called the degree of the vertex. Note that in Fig. 2-4 there are vertices of odd degree. Thus, we are led to assert and prove the following theorem.

Theorem 2-3 A map M can be two-colored if and only if every vertex of $U(M)$ has even degree ≥ 2.

• • • • •

Figure 2-3　　　　　　　　**Figure 2-4**

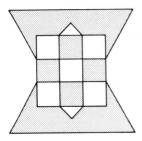

Figure 2-5

PROOF There are several proofs of this theorem. Here is one that is motivationally simple and which the reader can work out for himself. First, $U(M)$ cannot have an odd-degree vertex for then the countries which ring around this vertex would require more than two colors.

We prove the converse in three steps.

1. Suppose we generate a map in the entire plane by drawing straight lines extended indefinitely in both directions. We draw a line and color the half plane on one side with one color and the other half plane with another color. We can color such a map generated by n lines with two colors if, wherever we add a new line, we reverse all the colors (e.g., turn black to white and white to black) in the regions lying on only one side of the line. By induction on the number of lines, this can be shown to lead to a proper coloring of the map.
2. The same applies to a map generated by drawing circles in the plane and treating the interior and exterior of a circle as the two sides of a straight line.
3. We now return to our map with even (≥ 2)- degree vertices. Start a tour from any vertex traveling along the edges. This tour will generate a circuit somewhere once it returns to some vertex, as it must. This is because there is a finite number of vertices, and eventually the tour must double back on itself. Remove the edges of the circuit thus generated. No vertex has increased its degree. The graph left behind also has all its vertices of even degree and we repeat the process, decomposing the graph into circuits and chains. We then treat the circuits as circles and chains as lines to two-color the map.

Turning to the two-coloring of graphs we note that a circuit of odd length cannot be two-colored. Thus, any two-colorable graph cannot contain such a circuit. The next result says that the converse holds as well (see Fig. 2-6).

Theorem 2-4 The vertices of a graph G can be two-colored if and only if G contains no odd-length circuit.

PROOF To see the converse, pick some vertex v and color it red. Now color all vertices adjacent to v blue. Continue in this way until all vertices are colored.

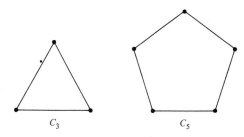

C_3 C_5 **Figure 2-6**

Everything would be fine if we could now assert that any two vertices w, x colored the same are not adjacent. However, the picture might look like that shown in Fig. 2-7, in which w and x are adjacent and colored the same.

The chain of edges from v to w and the chain from v to x, together with an edge from w to x, will be denoted by U. If U is a circuit, it certainly has odd length since the chains from v to w and v to x have the same parity because w and x are colored the same and the edge from w to x makes the length odd. In any case, it is easy to see that U must contain at least one circuit of odd length.

Here is a somewhat different way to do the above proof. If $a, b \in V(G)$, define the distance from a to b, $d(a, b)$, to be the length of the shortest chain from a to b. Now suppose G has no odd-length circuits. Color a point w red if $d(v, w)$ is even and blue if $d(v, w)$ is odd. (Note that $d(v, v) = 0$ so v is colored red.) If w and x are colored the same and w is adjacent to x, then $d(v, w) = d(v, x)$, for if $d(v, w) < d(v, x)$, then the shortest chain from v to w followed by the edge $[w, x]$ constitutes a chain from v to x of length $1 + d(v, w)$. But $d(v, x) \geq 2 + d(v, w)$ since x and w are colored the same. Let $\omega = (v = v_0, v_1, \ldots, v_t = w)$ and $\zeta = (v = v'_0, \ldots, v'_t = x)$ be the shortest chains from v to w and v to x, respectively (so $t = d(v, w) = d(v, x)$). If $y = v_r = v'_s$ is the last point of ω which the two chains have in common, then $r = s$ by minimality of both chains with respect to length. The same argument shows that y must also be the last point of ζ which the two chains have in common. Hence, the following sequence of vertices is a circuit of length $1 + 2(t - r)$:

$$(v_r, v_{r+1}, \ldots, v_t = w, \quad x = v'_t, v'_{t-1}, \ldots, v'_s = v_r)$$

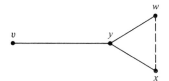

v y w x **Figure 2-7**

By applying this result to the dual graph of a map, we find that a map M is two-colorable if and only if $D(M)$ has no odd circuits. It is easy to see that each vertex v of $U(M)$ contributes a circuit z_v in $D(M)$ with length $= \deg(v)$. Later, when we shall analyze the structure of $D(M)$ more carefully, we will see that if all the circuits Z_v are even, then all the circuits in $D(M)$ are even; this implies Theorem 2-3.

We have now reached three-colorability and have just about run out of nice characterizations. Call a map *cubic* if each vertex has degree 3.

Theorem 2-5 A cubic map is three-colorable if and only if each region is bounded by an even number of sides.

PROOF First, let us prove that three-colorability of a cubic map implies that every region is bounded by an even number of edges. Suppose that there is a region R_1 with an odd number of edges. We can color R_1 with C_1 and its surrounding regions by colors C_2 and C_3 alternately, but then we have to color the last of the neighboring regions with a fourth color as it has a boundary with three regions with three different colors. This contradicts three-colorability and completes the first part of the proof.

Now it must be shown that if every region in a cubic map has an even number of edges, then the map can be three-colored. The proof is by induction on the number of regions r of the map.

When $r = 3$, the implication is trivial.

When $r = 4$, it can be shown easily by exhaustion of all possible cases that it can be colored with three colors (see Fig. 2-8).

Now assume that the theorem is true when there are $r - 1$ or r regions. Let M be a cubic map with $r + 1$ regions and let every region have an even number of edges. We shall soon see that in any map there must exist a region R with at most five edges. Since every region in G has an even number of edges, R has two or four edges.

Case 1 (Fig. 2-9) R has two edges, say common with regions R_1 and R_2. If we delete the edge between R and R_1, we obtain a map M' which has r regions where each region still has an even number of edges. (In M' every region has the same number of edges as in M, except that region $R'_1 = R_1 + R$ has two less edges than R_1 and, similarly, R'_2 has two less than R_2. We assume

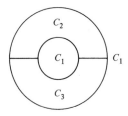

Figure 2-8

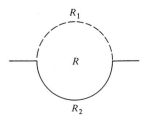

Figure 2-9

that the edges incident to a vertex of degree 2 are counted as one edge.) Therefore, M' is properly three-colorable, say with colors C_1, C_2, and C_3. If R'_1 has color C_1 and R'_2 has color C_2 in M', then restore the deleted edge and color R_1, R_2, and R with colors C_1, C_2, and C_3, respectively, keeping the colors of the other regions the same. Then M is obviously three-colored.

Case 2 (Fig. 2-10) R has four edges, say with regions R_1, R_2, R_3, and R_4, in that clockwise order. Suppose regions R_1 and R_3 have a common edge or coincide. Then, obviously, R_2 and R_4 can neither coincide nor have a common edge. Removing the edges between R and R_2 and between R and R_4, we thus obtain a map M' which has $r - 1$ regions; it can easily be seen, as in Fig. 2-10, that every region still has an even number of edges. Then M' can be three-colored with colors C_1, C_2, and C_3, by the hypothesis. It only remains to show that the regions R'_1 and R'_3 have the same color in M' (this is obvious if they coincide). Suppose the region $R' = R + R_2 + R_4$ has color C_1. Since there are an odd number of edges from v_a to v_b along that part of the boundary of R', the neighboring regions along these edges must have the alternating sequence of colors $C_2, C_3, C_2, \ldots, C_3, C_2$, as M' is three-colored. Then it is obvious that the regions R'_1 *and* R'_3 have the same color C_3. Restoring the deleted edges and coloring regions R, R_2, and R_4 with colors C_2, C_1, and C_1, respectively, while keeping the colors of the other regions the same as in M', we have a properly three-colored map M.

A map is *triangular* if every region has three sides. There is a good characterization of those triangular maps which can be three-colored.

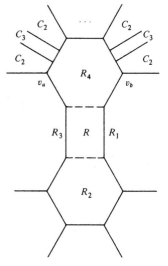

Figure 2-10

Theorem 2-6 A triangular map M can be three-colored if and only if $U(M) \neq K_4$.

REMARK This theorem is a direct application of Brooks' Theorem (given later) to the dual graph $D(M)$. Note that the dual of Theorem 2-5 states that the *vertices* of any triangular graph M can be three-colored *if* and *only if* every vertex has even degree.

The following theorem, due to Grötzsch (1958), is one of the few general results on three-coloring planar graphs.

Theorem 2-7 If G is a planar graph which contains no cycle of length 3, then G can be three-colored.

2-3 THE CHROMATIC NUMBER

We ran out of ways of characterizing graphs which can be colored with a given number of colors. Now we turn the problem around and ask: "What is the minimum number of colors required for coloring the vertices of any graph?" Can the answer be given in terms of the properties of the particular graph? Can it be given for a general class of graphs? It turns out that the best characterization of the minimum number of colors is given in terms of the surfaces in which graphs are embedded, as we shall see in subsequent sections of this chapter.

In this section we give a few general results about coloring. We find that one can inductively construct a k-coloring of a graph G provided that every subgraph of G contains some vertex of degree less than or equal to $k - 1$, and this result will provide the key to the coloring problems we will consider in later sections.

The *chromatic number* $\chi(G)$ of a graph G is the smallest integer k for which G can be k-colored; G is called k-chromatic. The same definitions apply to maps. For any graph G, $\Delta(G)$ denotes the maximum degree at any vertex and $\delta(G)$ the minimum degree. The following theorem gives a bound on $\chi(G)$ in terms of $\Delta(G)$.

Theorem 2-8 For any graph G, $\chi(G) \leq 1 + \Delta(G)$.

PROOF We induct on the number n of vertices. If G has one vertex, the result is certainly true. Assume that it also holds if G has $n - 1$ vertices ($n \geq 2$) and let G be any graph with n vertices. Suppose the degree of v is $\Delta(G)$. Then, by the inductive hypothesis, $\chi(G') \leq 1 + \Delta(G') \leq 1 + \Delta(G)$, where $G' = G - v$ is obtained from G by removing v and its incident edges. But now we can color v using one of the $1 + \Delta(G)$ colors not used by the $\Delta(G)$ vertices of G which are adjacent to v.

That the result is best possible follows from $\chi(K_{n+1}) = n + 1$ while $\Delta(K_{n+1}) = n$. However, in (almost) every other case we can improve the preceding theorem by 1, as the following result of Brooks (1941), proved in Chapter 6 (page 137), shows.

Theorem 2-9 Let G be a connected graph and let $n = \Delta(G)$. Then $\chi(G) = n$ unless either

(a) G is a complete graph of order $n + 1$, or

(b) $n = 2$ and G is a cycle of odd length.

The bound given in Brooks' theorem, while a very useful general result, doesn't come very close to the actual chromatic number known in many important cases. For example, we will show that $\chi(G) \le 5$ for any planar graph G, even though there exist planar graphs with arbitrarily large maximum degrees. In fact, if G is any nontrivial connected graph and r is a positive integer, we can attach a star-shaped "barnacle" (as indicated in Fig. 2-11) to obtain a $\tilde{G}$ with $\Delta(\tilde{G}) \ge r$ and $\chi(\tilde{G}) = \chi(G)$. If G is planar, so is $\tilde{G}$.

If one reexamines the proof of Theorem 2-8, one can see that only the existence of some vertex of specified degree is required. This suggests a definition. Let $sw(G)$ be the max $\{\delta(H) \mid H \subset G\}$. ($H$ is a *subgraph* of G, $H \subset G$, means $V(H) \subset V(G)$ and $[v, w]$ in H implies $[v, w]$ in G.) The following result of Szekeres and Wilf (1968) is now proved exactly like Theorem 2-8.

Theorem 2-10 $\chi(G) \le 1 + sw(G)$.

PROOF We actually prove that $\chi(G') \le 1 + sw(G)$ for any subgraph G' of G. Let n' be the number of vertices in G'; we use induction on n'. If $n' = 1$, the result

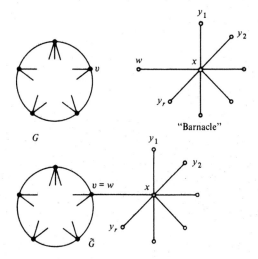

Figure 2-11

is clear. Suppose it holds for $n' = k \geq 1$ and let G' have $k + 1$ vertices. By definition of $\text{sw}(G)$, there is a vertex v in G' with degree at most $\text{sw}(G)$. By the inductive hypothesis, $G' - v$ can be colored with at most $1 + \text{sw}(G)$ colors and the neighbors of v can use up at most $\text{sw}(G)$ of these colors, leaving one free for v. Hence, $\chi(G') \leq 1 + \text{sw}(G)$.

But what does $\text{sw}(G)$ mean? Clearly, $\text{sw}(G) = 0$ if and only if G has no edges. If every vertex of graph G has degree > 1, then G contains a circuit. Hence, a nontrivial tree must have at least one vertex of degree 1 (called an *endpoint* of the tree), and it is an easy consequence that $\text{sw}(G) = 1$ if G is a nontrivial tree. On the other hand, if G is nontrivial and not a tree, then G contains a circuit C and $\text{sw}(G) \geq \delta(C) = 2$. Hence a connected nontrivial graph G satisfies $\text{sw}(G) = 1$ if and only if G is a tree. Thus, by Theorem 2-10 any tree can be two-colored. Of course, we could also derive this result from Theorem 2-4 since a tree certainly has no odd circuits.

The condition $\text{sw}(G) = 2$ is not very restrictive, for if G is any graph, we may form a new graph G^* by introducing a new vertex of degree 2 into the middle of each edge of G (see Fig. 2-12). The significance of this construction is that, without defining the term precisely, it is plain that G is planar if and only if G^* is planar.

Theorem 2-11 Let G be a planar graph. Then $\delta(G) \leq 5$.

Corollary 2-1 If G is planar, then $\text{sw}(G) \leq 5$.

Corollary 2-2 Any map has a region with at most five sides.

PROOF Let $d = d(G) = (1/n) \Sigma \deg(v)$ be the *average degree* of a vertex of G. Since $2m = \Sigma \deg(v)$, v a vertex of G, we have $dn = 2m$. But $m \leq 3n - 6$ so $d \leq 6 - (12/n) < 6$. Hence, $\deg(v) \leq 5$ for some vertex v in G; otherwise, the average degree would be at least 6.

Corollary 2-1 follows since every subgraph of a planar graph is planar. Hence, $\delta(H) \leq 5$ for every subgraph H of G. Corollary 2-2 is derived by applying the theorem to the dual graph of the map.

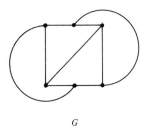

G

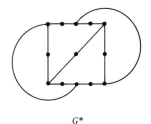

G^*

Figure 2-12

The next theorem, which is an immediate consequence of Theorems 2-10 and 2-11, is actually redundant since we can prove that every planar graph can be five-colored. However, we state it anyway since many coloring results are analogous to this easy theorem rather than to the harder five-color theorem.

Theorem 2-12 (six-color theorem) Let G be a planar graph. Then G can be six-colored.

Corollary 2-3 Any map can be six-colored.

It requires a much more careful argument to show that any planar graph can be five-colored. The original method was found by Kempe (1879) who thought he had proved the 4CC. Heawood (1890) discovered Kempe's error and went on to show how the argument could be used to prove the five-color theorem. This is the first proof we give of the result. A second proof (of a more topological nature) is also given.

This theorem is part of the reason why the 4CC is so tantalizing. We knew the correct answer was either 4 or 5, but we did not know which! It ought to be mentioned here that many fine mathematicians actually believed that the answer was 5, not 4; and some referred to a five-color conjecture, which is the assertion that some planar five-chromatic graph exists.

Theorem 2-13 (five-color theorem) Every planar graph can be five-colored.

Corollary 2-4 Every map can be five-colored.

FIRST PROOF We proceed by induction on the number n of vertices. When $n \leq 5$, the result is trivial. Suppose it holds for $n - 1$, and let G be an arbitrary planar graph with n vertices. By Theorem 2-11, choose a vertex v of degree ≤ 5 and let $G' = G - v$. Now G' has $n - 1$ vertices and, hence, by the inductive hypothesis, G' has a five-coloring C'. If v has fewer than five adjacent vertices in G, then C' can be extended to a five-coloring C of G by assigning to v one of the colors not used by any of its neighbors. Hence, we may suppose that v has precisely five neighbors, say v_1, v_2, v_3, v_4, v_5, listed in clockwise order, and that the five colors assigned to these vertices are all distinct (see Fig. 2-13). If x and y

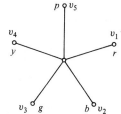

Figure 2-13

are colors, let us denote by $G'(x, y)$ the *full subgraph* of G' *induced* by all vertices colored x or y. (This just means that whenever two vertices of $G'(x, y)$ are adjacent in G, they are adjacent in $G'(x, y)$.) A component of $G'(x, y)$ is also called an x–y component.

We shall now show how to alter the coloring C' so that all five colors used by the neighbors of v are not distinct, thereby enabling us to color v with one of the leftover colors.

Consider v_1 and v_3; either they belong to the same r–g component or they do not. If not, then we may change the coloring C' by reversing the colors r and g in the r–g component containing v_1. Then v_1 and v_3 are both colored g, while the colors of v_2, v_4, and v_5 do not change.

If v_1 and v_3 do belong to the same r–g component, then there is a chain in $G'(r, g)$ from v_1 to v_3, and this chain together with the two edges v_1, v and v_3, v in G form a circuit which must separate v_2 from v_4. Thus, there cannot be a chain in $G'(b, y)$ from v_2 to v_4 so the b–y component of v_2 does not include v_4. Therefore, we may reverse the colors b and y in this component, so that both v_2 and v_4 are colored y. This type of argument is called a Kempe-chain argument, and the process by which we modified C' is called Kempe chaining.

SECOND PROOF Once again, by induction, we suppose that every planar graph with fewer than n vertices can be five-colored and let G be planar with n vertices. Choose a vertex v in G with degree at most 5. As before, we may assume that $\deg(v) = 5$. If every pair of vertices adjacent to v were themselves adjacent, G would contain a subgraph isomorphic to K_5 (in fact, it would contain a subgraph isomorphic to K_6!) and this is impossible. Let v_1 and v_2 be vertices adjacent to v but not adjacent to each other. Let e be an edge joining v_1, and v and f an edge joining v_2 and v; delete every other edge incident with v. Contracting the edges e and f produces a new graph $\bar{G}$ in which v_1 and v_2 are identified with v to produce a new vertex $\bar{v}$ and in which no adjacencies between vertices other than v are added or removed (see Fig. 2-14). Moreover, $\bar{G}$ is planar and has fewer than n vertices so $\bar{G}$ may be five-colored. But a five-coloring of $\bar{G}$ yields a five-coloring of $G - v$ in which the two vertices v_1 and v_2 receive the same color. As before, this suffices to five-color G.

Note that the second proof only requires that K_6 be nonplanar.

When Heawood (1890) proved the five-color theorem by salvaging an attempted proof by Kempe of the 4CC, he gave a counter-example to show that Kempe's argument was fallacious in addition to adapting the method to prove the sufficiency of five colors.

Heawood's counter-example is directed at Kempe's chain-coloring reversals. He is not concerned with whether one can, by a judicious choice, recolor some of the vertices (see Fig. 2-15). This example has 25 vertices and is known to be four-colorable by existing theory.

Using the inductive argument on the number of vertices n, assume that every

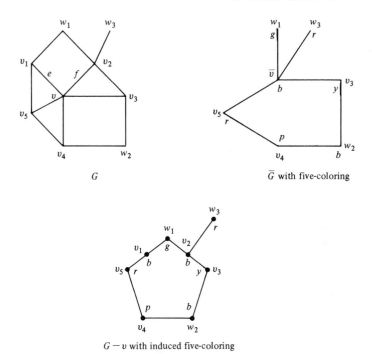

G

$\overline{G}$ with five-coloring

$G - v$ with induced five-coloring

Figure 2-14

planar graph on < 25 vertices is four-colorable. Consider a graph on 25 vertices and remove a vertex v (which has five neighbors) and its connecting edges,, and four-color the resulting graph on 24 vertices. Suppose the coloring is as shown. Reinstate v and attempt to color the resulting graph.

There is a b–g chain from 2 to 4. There is also a b–y chain from 2 to 5. Reversal of colors on either chain will not free a color for v. This leaves r in two places. Now there is no r–g chain from 1 to 4. Therefore, one can reverse r to g in the r–g chain starting at 1. But the other r at 3 must also be turned to g or to y to obtain the spare color for v. This is not possible because 4, which has color g, is adjacent to 3 which will become colored with g. On the other hand, if we reverse colors on the r–y chain starting at 3, the two vertices of the outer triangle which are connected by an edge would both be assigned r by the r–g and r–y reversals starting at 1 and at 3, respectively, contradicting proper coloring. Thus, one cannot replace r by g at both 1 and 3, nor by g at 1 and by y at 3. Note that at 1, r cannot be turned to y because it is adjacent to a y at 5. Heawood wrote, "Unfortunately, it is conceivable that though either transposition would remove an r both may not remove both r's." (It is clear that reversal of colors on part of the y–r chain starting at 5 followed by a reversal on the r–g chain starting at 1, frees the color y for v, but this does not justify Kempe's argument.) See also Saaty (1967).

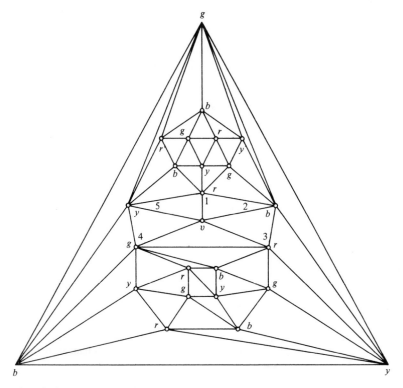

Figure 2-15

We have seen five- and six-color theorems. For a seven-color theorem, we will consider a graph G on the *torus*, which is the surface shaped like a doughnut. The analogue of Euler's formula on the torus is the formula $n - m + r = 0$, from which we conclude that if G has n vertices and m edges, then $m \leq 3n$. Thus, the average degree d is at most six, and so, arguing as before, $\mathrm{sw}(G) \leq 6$. Therefore, we can conclude $\chi(G) \leq 7$ when G is torodial. The next result says that this is the best possible.

Theorem 2-14 If G is toroidal, $\chi(G) \leq 7$ and there is a toroidal graph G with $\chi(G) = 7$; namely, $G = K_7$.

Corollary 2-5 Any map on the torus can be seven-colored and some toroidal map requires seven colors.

PROOF All that remains is to show that K_7 is toroidal. Instead, we exhibit a map on the torus consisting of seven mutually adjacent regions (see Fig. 2-16).

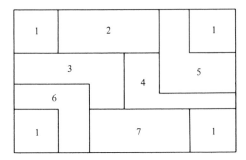

Figure 2-16

To make this rectangular figure into a torus, first imagine that the upper and lower edges of this rectangle are bent toward each other until they touch. In the resulting figure, which looks like a cylinder, regions 2 and 7, for example, have become adjacent. The four regions labeled 1 have been fused into two regions, one on the left and one on the right. Now bring the two ends of the cylinder together to create a torus. (See Sec. 2-4 for a more complete discussion.)

The *empire problem* asks: "Given r a nonnegative integer, what is the minimum number N_r of colors needed to color a map in which some countries have up to r colonies?" Of course, a country and its colonies should be colored the same, and no pair of colonies or of country and colony are adjacent. If $r = 0$, this is just the four-color problem. However, if $r = 1$, the problem was solved in 1939.

Theorem 2-15 For $r \geq 0$, $N_r \leq 6(r + 1)$.

PROOF It is convenient to transform the problem into a graph-coloring question. If M is the map we want to color, let G be the simple graph obtained from $D(M)$ by identifying all the vertices corresponding to a country and its colonies (see Fig. 2-17) and then deleting any resulting loops or multiple edges.

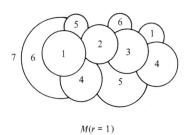

$M(r = 1)$

$G = K_7$

Figure 2-17

It is easy to see that if G has m edges and n vertices, then $m \leq 3(r + 1)$ $(n - 2)$. Hence, the average degree d is less than $6(r + 1)$ so $\mathrm{sw}(G) \leq 6(r + 1) - 1$. Therefore, by Theorem 2-10, $\chi(G) \leq 6(r + 1)$.

When $r = 1$, Franklin (1939) gives the following example (see Fig. 2-18) of a map M so that the associated graph $G = K_{12}$. Hence, $N_1 = 12$.

For $r > 1$, as far as we know, no one has constructed the requisite maps to show $N_r = 6(r + 1)$.

If we move the coloring problem for empires into the latter half of the twentieth century, we might find ourselves confronted with the following question. Assume that the moon is divided into regions, each of which belongs to a nation on earth; assume further that each country has at most one lunar colony. What is the minimum number N of colors needed to color the earth and moon simultaneously so that each country and its lunar colony get the same color?

The graphs G corresponding to such "maps" are characterized by the property that they have *thickness* $\theta(G) = 2$. (The thickness $\theta(G) = t$ of any graph is the minimum number of subgraphs $G_1, \ldots, G_t$ of G such that $G = G_1 \cup \ldots \cup G_t$ and each G_i is planar.) If $\theta(G) = 2$, it is easy to see that $\mathrm{sw}(G) \leq 11$; the argument is just like that for the empire problem with $r = 1$. However, that is

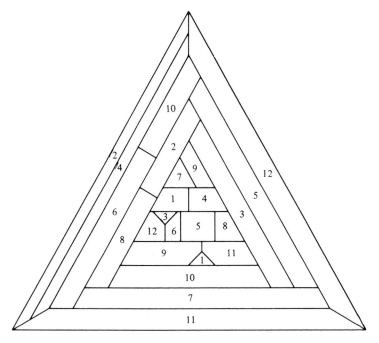

Figure 2-18

where the similarity ends. For example, K_{12} does *not* have thickness 2. In fact, the greatest order of any complete graph of thickness 2 is only 8. If N is the largest number of colors needed to color any graph of thickness 2, we have (by Theorem 2-10) the following theorem.

Theorem 2-16 $8 \le N \le 12$.

Recently Sulanke (private communication 1973) has given an example of a graph of thickness 2 requiring nine colors. The graph consists of K_6 and C_5 with each vertex in K_6 joined by edges to each vertex in C_5. This shows that actually $N \ge 9$.

2-4 SURFACES AND THE HEAWOOD NUMBER

This section should make very interesting, perhaps exciting, reading because of its content and because it not only gives a bound (the Heawood number) on the number of colors needed to color any map on any surface except for the plane but also shows that this number is both necessary and sufficient, thus making the bound exact or sharp for such surfaces. Unfortunately, for the plane, the derivation of the bound is illegitimate. By other means we have obtained a bound of five and we know that four colors are necessary. The next chapter will be concerned with the proof that the bound for the plane is actually four. That is the four-color theorem.

Trying to discuss the four-color problem without including at least an intuitive description of topology would be like trying to put up a building without a framework of girders, for underlying all of our work on the 4CC is the essential *topological* restriction that all graphs be planar. So let us now proceed to give an intuitive idea of topology.

In geometry there are two main notions of equivalence between figures in euclidean space: congruence and similarity. Two figures are said to be *congruent* if they are identical except for their position in the containing space. Thus, two figures are congruent if we can place one figure on the other so that they coincide. Two figures are *similar* if they are identical except for their position and size. Thus, they are similar if, after changing the size of one, we can place it on the other.

In topology the main equivalence relation is homeomorphism. Two figures are said to be homeomorphic if each can be transformed into the other by a "continuous deformation." Thus, two figures are homeomorphic if, after deforming one of the figures, we can place it on the other, and vice versa. For example, the circle and the ellipse are homeomorphic since a circle can be continuously deformed (or stretched) into an ellipse and an ellipse can be deformed into a circle.

We must not, of course, limit ourselves to only two or three dimensions in performing these deformations. For example, two disjoint circles and two linked circles are certainly homeomorphic figures, but the deformation requires at least four dimensions. Let us give another definition of homeomorphism which does not depend on deformation. Two figures are homeomorphic if there is a one-to-one correspondence between their points so that two points are "close" in one, if and only if the corresponding points in the other figure are also "close." For example, we can represent the points of a circle of radius 1 by

$$\{(\cos\theta, \sin\theta) \,|\, 0 \le \theta \le 2\pi\}$$

and the points of an ellipse with major axis $2a$ and minor axis $2b$ by

$$\{(a\cos\theta, b\sin\theta) \,|\, 0 \le \theta \le 2\pi\}$$

There is an obvious one-to-one correspondence between their points $(\cos\theta, \sin\theta) \leftrightarrows (a\cos\theta, b\sin\theta)$, and this correspondence preserves "closeness."

In general, it is not always so easy to write down such a one-to-one correspondence, or *homeomorphism*, between two homeomorphic figures. It is even harder to show that no such correspondence exists, even when the figures are clearly not homeomorphic. For example, one can certainly believe that the circle and the figure 8 are not homeomorphic since it is intuitively impossible to continuously deform each figure into the other. The reason is that these two geometric objects have different *intrinsic* properties (an intrinsic property is a property of the figure itself and not of its relationship to the surrounding space). For example, if a circle is cut at any point, it remains in one "connected" piece, while if a figure 8 is cut at the junction of its upper and lower circles, it breaks into two connected pieces. Thus, there is a *qualitative* difference between the circle and the figure 8, but only a *quantitative* difference between the circle and the ellipse. In other words, if we had no way to count or to accurately measure distance, we could not distinguish between the circle and the ellipse, but we could still differentiate the circle from the figure 8 by a simple experiment with a pair of scissors.

Another example of an intrinsic, or topological, property is the notion of connectedness. A figure is *connected* if any two points in the figure can be joined by a continuous curve. Thus, one circle and two disjoint circles cannot be homeomorphic since the first figure is connected while the second is not.

Now we want to relate topology to graphs. Suppose that G is a graph. Take a collection of points in three-dimensional euclidean space R^3, one for each vertex of G. Join a pair of points by a simple arc, which passes through no other point, if and only if the corresponding two vertices of G are adjacent. The simple arcs are allowed to intersect each other only at their endpoints. The resulting geometric figure is called a *topological realization* of the graph G.

Any two topological realizations of G are homeomorphic since corresponding arcs and their endpoints can be put in one-to-one correspondence. Thus, we will speak of the topological realization $|G|$ of G. Many properties of a graph are reflected in its realization. For example, G is connected if and only if $|G|$ is connected.

It is not obvious that every graph has a topological realization, but, in fact, if the graph is simple, we can even construct a realization in which all of the arcs are straight-line segments. Choose a collection of points in R^3, one for each vertex of the graph, so that no three points lie on the same line and no four points lie in the same plane. (This is called choosing the points in *general position*.)To do it, start with any two points p_1 and p_2. Take a third point p_3 not on the line $L(p_1, p_2)$ determined by p_1 and p_2. Now select a fourth point p_4 not lying in the plane $P(p_1, p_2, p_3)$ determined by the first three points. Thus, p_4 is not on the lines $L(p_1, p_2)$, $L(p_1, p_3)$, or $L(p_2, p_3)$. Take p_5 so that it does not lie in any of the four planes $P(p_1, p_2, p_3)$, $P(p_1, p_2, p_4)$, $P(p_2, p_3, p_4)$, $P(p_1, p_3, p_4)$ determined by $\{p_1, p_2, p_3, p_4\}$, and so forth.

Now take the chosen points and join them by straight-line segments when the corresponding vertices are adjacent. If a line segment passed through any point other than its endpoints, we would have three points on a line, which is impossible. If two line segments intersected anywhere except at an endpoint, then the four endpoints of these two line segments would all lie on the same plane, which is again impossible. Thus, we have constructed a realization for any simple graph. If the graph has loops or parallel edges, we can add them to the realization, but not by straight-line segments.

Suppose that two graphs G and G' have isomorphic subdivisions. Then $|G|$ and $|G'|$ are homeomorphic since the realization of an edge or path is just a simple arc. Conversely, if $|G|$ and $|G'|$ are homeomorphic, then there is a one-to-one correspondence between paths in G and in G', and so, by introducing sufficiently many new vertices of degree 2, we make these paths identical. Hence G and G' have isomorphic subdivisions. Thus, G and G' are homeomorphic in the graph-theoretic sense if and only if $|G|$ and $|G'|$ are homeomorphic in the topological sense.

Realizations allow us to make more precise our previous definitions of planarity and of maps. A graph is planar if it can be realized in the plane, i.e., if we can choose the points and simple arcs of a realization of the graph to lie in the plane. (In fact, G is planar if and only if $|G|$ can be embedded in the plane.) A map is then just a particular planar realization or embedding of some graph.

What properties do the sphere, the plane, and the torus have in common? Given any point x there is a (small) neighborhood of x which is indistinguishable from, or, as we prefer to say, homeomorphic to, the euclidean plane. We call such a neighborhood euclidean and refer to a point with such a neighborhood as euclidean. In the sphere, obviously, *every* point is euclidean. This is also true for the euclidean plane R^2, but R^2 differs from the examples in that they are bounded while it is unbounded. Given an embedding of the sphere or torus in R^3, there must exist some finite bound B such that any point in the embedded sphere or torus lies within distance B of the origin.

To see that an arbitrary point x on the torus is also euclidean, consider a corresponding point x' in the square. That is, x' is one of the points of the square which becomes x after the boundary of the square is identified in the appropriate way (see Fig. 2-16 and the remarks following it). If x' belongs to the interior of the square, then x' is certainly euclidean, and any small euclidean neighborhood of x', lying entirely within the interior of the square, is undisturbed by the identification process and, hence, forms a euclidean neighborhood of x. (See Fig. 2-19; the edges with two arrows are identified, as are those with one arrow. Edges are identified so that the direction of the arrow is preserved.)

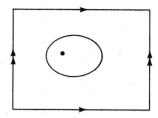

Figure 2-19

On the other hand, if x' lies on the boundary of the square except the corners, it is paired off with another point x'', each of which corresponds to x. Both x' and x'' have neighborhoods U' and U'' which are half discs. Under the identification, U' and U'' form a subset $\overline{U}$ containing a euclidean neighborhood of x (see Fig. 2-20). The same idea works for the four corner points as well.

Let us define a *surface* to be any *bounded figure* in which every point is euclidean. Of course, we have to give meaning to the terms "figure" and "bounded."

A *figure* is any subset X of four-dimensional euclidean space R^4; a figure is *bounded* if, no matter how it is embedded in R^4, it lies within some ball of finite radius about the origin. Note that we need R^4, not just R^3, to accommodate surfaces with twists like the Klein bottle (see page 42).

Let us now see how to construct surfaces. For a surface S and $X \subset S$, write $\mathring{X}$ for the set of euclidean points in X; $\mathring{X}$ is called the *interior* of X. If S^2 is the sphere, choose two disjoint subsets A and B which are homeomorphic to the closed unit disc and remove the interiors $\mathring{A}$ of A and $\mathring{B}$ of B, leaving only their two bounding circles C_A and C_B (see Fig. 2-21).

The points lying on C_A and C_B in $S^2 - (\mathring{A} \cup \mathring{B})$ are not euclidean (their small neighborhoods look like the euclidean half plane, not the whole plane). Now take a cylinder V with bounding circles C'_A and C'_B and form a new figure from $S^2 - (\mathring{A} \cup \mathring{B})$ and V by identifying C'_A with C_A and C'_B with C_B (see Fig. 2-22). Call the results S_1. We say that S_1 is obtained from the sphere by attaching a *handle*. In fact, it is not difficult to see that S_1 is homeomorphic to the torus. Note that all we know about the sphere is that it is possible to find disjoint closed discs. But this is true of any surface, and so we can attach a handle to any surface.

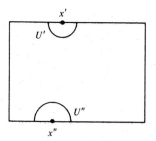

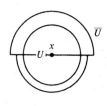

Figure 2-20

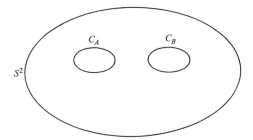

Figure 2-21

There is another way in which we can alter a surface. First, we need a brief digression to describe the Möbius strip. This figure is obtained, like the cylinder, by identifying one pair of opposite sides in a square, but this time in reverse order (see Fig. 2-23).

In the cylinder, the other pair of opposite sides become the two bounding circles. However, in the Möbius strip, these two sides become a *single* circle called the *boundary* of the Möbius strip. Then the Möbius strip is obtained by taking the two ends of a strip of paper and glueing them together after giving one end a half-twist (see Fig. 2-24). Note that the boundary wraps around twice.

If we deform the Möbius strip so that its boundary lies in a single plane, the result, which cannot be embedded in R^3, is called a crosscap (see Fig. 2-25).

Now if S is any surface, we can attach a crosscap to S by removing the interior $\overset{\circ}{A}$ of some closed disc A in S and then identifying the boundary of a Möbius strip with the bounding circle C_A of A (see Fig. 2-26). If S is the sphere, the resulting surface turns out to be the projective plane, called U_1, below.

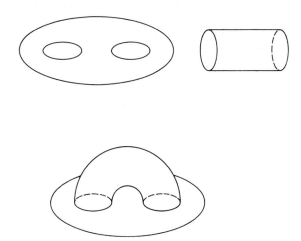

Figure 2-22

Figure 2-23

The following theorem classifies surfaces and shows that they can all be obtained from the sphere by attaching handles and crosscaps. Its proof would take us too far afield; we refer the interested reader to Cairns (1961) or Fréchet and Fan (1967).

Theorem 2-17 Let S be any surface. Then S is homeomorphic to one and only one of the following:

(a) S_k, the surface obtained from the sphere by attaching $k \geq 0$ handles (by convention S_0 means the sphere);

(b) U_k, the surface obtained from the sphere by attaching $k - 1 \geq 0$ handles and one crosscap;

(c) S'_k, the surface obtained from the sphere by attaching $k - 1 \geq 0$ handles and two crosscaps.

The surfaces U_k and S'_k, $k \geq 1$, are called *nonorientable*, while S_k is called the *orientable* surface of *genus k*. Note that U_1 is simply the projective plane and S'_1 is the Klein bottle (change sense of any arrow in Fig. 2-19).

If S is any surface and G is any graph, we write $G \subset S$ to mean that G is embedded or can be embedded in S. (Thus, $G \subset S_0$ if and only if G is planar.) Call $X \subset S$ *open* if $X = \mathring{X}$. It is not hard to show that, for $G \subset S$, $S-G$ is open. Call $X \subset S$ connected if any two points in X can be joined by a continuous arc lying entirely within X. Any $X \subset S$ can be written as the union of *connected components X_i*, each of which is a maximal connected subset of S. If $X \subset S$ is open, then since every point in X is euclidean, it can be shown that the connected components X_i of X are also open.

Now suppose $G \subset S$. By the above, the connected components of $S-G$ are open; they are called the *regions* of the embedding. If every region is homeo-

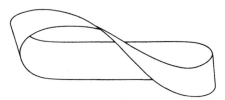

Figure 2-24

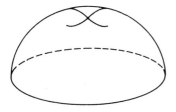

Figure 2-25

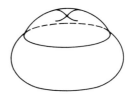

Figure 2-26

morphic to the interior of the closed unit disc, then the embedding is called a *two-cell embedding* (Youngs, 1963). If G has a two-cell embedding in S, we write $G < S$. Any embedding of a connected graph in the sphere is a two-cell embedding, but this need not be true for other surfaces. If S is any surface, we define the *Euler characteristic* $e(S)$ by the formula

$$e(S) = 2 - 2k - j$$

where $k \geq 0$ is the number of handles and $j = 0, 1,$ or 2 is the number of crosscaps. Thus, $e(S_0) = 2$, $e(U_1) = 1$, and $e(S_1) = 0 = e(S_1')$.

We can now state Euler's formula for an arbitrary surface S.

Theorem 2-18 If G is a connected graph with n vertices and m edges and $G < S$ with r regions, then $n - m + r = e(S)$.

Note that the restriction $G < S$, rather than $G \subset S$, is really needed. For $C_3 \subset S_1$; if Euler's formula held for the embedding, we would be forced to conclude that $C_3 \subset S_1$ produced no regions.

We can use Theorem 2-18 to produce a necessary condition for G to have a two-cell embedding in S.

Theorem 2-19 Let G be a connected simple graph with n vertices and m edges. If $G < S$, then $m \leq 3 (n - e(S))$.

The next theorem relates embeddings and two-cell embeddings.

Theorem 2-20 Supose $G \subset S$ but the embedding is not two-cell. Then there is a surface $\overline{S}$ such that $G < \overline{S}$ and $e(\overline{S}) > e(S)$..

The idea of the proof, due to König (1936) but carried out carefully by Youngs (1963), is to just cut off any handle or crosscaps occurring entirely within any one region of $S-G$. The "stumps" are then sealed off with discs (see Fig. 2-27).

Theorem 2-20 allows us to put Theorem 2-19 into a more useful form.

Corollary 2-6 If G is a simple graph with n vertices and m edges and $G \subset S$, then $m \leq 3 (n - e(S))$.

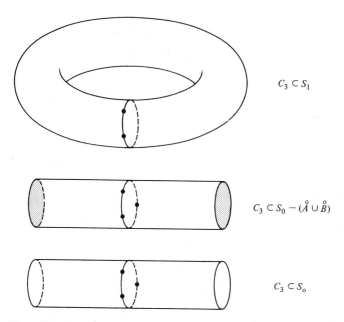

$C_3 \subset S_1$

$C_3 \subset S_0 - (\mathring{A} \cup \mathring{B})$

$C_3 \subset S_\circ$

Figure 2-27

If G is a simple graph with m edges and n vertices, recall that $d(G) = 2m/n$. It follows that:

Lemma 2-1 Let G be simple. If $G \subset S$ has n vertices, then

$$d(G) \leq 6 - \frac{6e(S)}{n}$$

For example, if $G \subset U_1$, then $d(G) \leq 6 - (6/n)$ and, thus, $sw(G) \leq 5$. Therefore, $\chi(G) \leq 6$ whenever $G \subset U_1$.

We are about to show that this method generalizes to all surfaces other than the sphere. First, consider the equation $x^2 - 7x + 6e(S) = 0$. There are two roots $\alpha(S)$ and $\alpha^*(S)$ given by

$$\alpha(S) = \frac{7 + \sqrt{49 - 24e(S)}}{2}$$

and

$$\alpha^*(S) = \frac{7 - \sqrt{49 - 24e(S)}}{2}$$

Set $H(S) = [\alpha(S)]$, the greatest integer in $\alpha(S)$. This is called the Heawood number of the surface S because of the following theorem which is due to Heawood (1890).

Theorem 2-21 Let S be a surface other than the sphere. Then $\chi(S) \leq H(S)$.

PROOF If S is the sphere, $H(S) = 4$. Unfortunately, the arguments do not apply in this case.

We have already proved this theorem for the projective plane since $H(U_1) = (7 + \sqrt{49 - 24})/2 = (7 + 5)/2 = 6$. To finish the proof, we need only show that if S is any surface other than the sphere or projective plane and $G \subset S$, then $sw(G) \leq H(S) - 1$. In fact, we derive a stronger result—that $d(G) \leq \alpha(S) - 1$. This suffices since the inequality $d(H) \leq \alpha(S) - 1$ holds for all subgraphs H of G. Hence, $\delta(H) \leq d(H) \leq \alpha(S) - 1$. But $\delta(H)$ is an integer so $\delta(H) \leq H(S) - 1$. Maximizing over all $H \subset G$ yields $sw(G) \leq H(S) - 1$.

Theorem 2-22 Let S be a surface other than S_0 or U_1 and let $G \subset S$ be a simple graph. Then $d(G) \leq \alpha(S) - 1$.

PROOF Suppose that G has n vertices. If $n \geq \alpha(S)$, then

$$d(G) \leq 6 - \frac{6e(S)}{n} \leq 6 - \frac{6e(S)}{\alpha(S)} = \alpha(S) - 1$$

The first inequality uses Lemma 2-1; the second inequality uses the fact that $e(S) \leq 0$, and the equality is a direct consequence of the definition of $\alpha(S)$. On the other hand, if $n < \alpha(S)$, then $d \leq n - 1 < \alpha(S) - 1$.

For S, any orientable surface other than the sphere, and for S, any non-orientable surface other than the Klein bottle, Ringel and Youngs (1968) and Ringel (1959), respectively, have shown $K_{H(S)} \subset S$. This implies that Theorem 2-21 is the best possible. Let $\chi(S) = \sup\{\chi(G) | G \subset S\}$.

Corollary 2-7 For $S \neq S_0$ or S_1', $\chi(S) = H(S)$.

For S_1', $H(S_1') = 7$, but Franklin (1934) showed that K_7 does *not* embed in S_1'. Since $K_6 \subset S_1'$, an argument exactly like that by which the five-color theorem was established shows the following theorem.

Theorem 2-23 $\chi(S_1') = 6$.

Thus, the problem of coloring graphs on any surface but the sphere is completely solved. In every case, the answer is given by the largest number of vertices in any complete graph which is embeddable in the surface. If this were true for the sphere, we would have proved the 4CC since K_4 is planar but K_5 is not.

There are two important questions we ought to answer now:

1. Does every graph embed in some surface?
2. Can we use information about the chromatic number of surfaces together with embedding graphs in surfaces to get at the chromatic number of the graphs?

The answer to the first question is "Yes," as we shall see; but the answer to the second is a resounding "No." Given any integer p there is a bipartite graph which is not embeddable in an orientable surface of genus less than p.

Theorem 2-24 Every graph G embeds in some orientable surface.

PROOF The idea of the proof is simple. Just draw the graph on the surface of the sphere allowing edges to cross one another. Now add a "thin" handle for each edge $e = [v, w]$ by attaching both ends of the handle near the vertices v and w and avoiding all the rest of the graph; the handles can be deformed in R^3 to miss one another. Of course, we do not really need a separate handle for each edge. For example, in Fig. 2-28, we have embedded K_5 in the sphere with one handle attached.

In view of Theorem 2-24, we can define the genus $\gamma(G)$ of a graph G to be the smallest integer p such that $G \subset S_p$. Because of Corollary 2-6, if G is simple and has n vertices and m edges, then $m \leq 3(n - e(S\gamma(G))) = 3(n - 2\gamma(G)) = 3n - 6 + 6\gamma(G)$. Similarly, if G has no triangles, then $m \leq 2n - 4 + 4\gamma(G)$. Thus, we have the following corollary.

Corollary 2-8 If G is simple and has n vertices and m edges, then $\gamma(G) \geq 1/6(m - 3n + 6)/6$ and, if G has no triangles, $\gamma(G) \geq 1/4(m - 2n + 4)/4$.

For example, if $G = K_{t,t}$ is the complete bipartite graph, then $n = 2t$, $m = t^2$, and so $\gamma(K_{t,t}) \geq \frac{1}{4}(t^2 - 4t + 4) = (t - 2)^2/2$. Since $\gamma(K_{t,t})$ is an integer, this implies $\gamma(K_{t,t}) \geq \left\{ \dfrac{(t - 2)^2}{2} \right\}$, where $\{x\}$ is the least integer $\geq x$. Thus, for any integer p, by taking t sufficiently large, the bipartite graph $K_{t,t}$ does not embed in any orientable surface of genus less than p. In fact, the inequality given above is actually an equality (Ringel, 1965).

Theorem 2-25 For all $s, t \geq 2$,

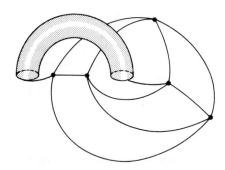

Figure 2-28

$$\gamma(K_{s,t}) = \left\{ \frac{(s-2)(t-2)}{4} \right\}$$

PROOF The proof consists of giving the requisite embeddings of the $K_{s,t}$.

A similar result for complete graphs is due to Ringel and Youngs (1968). For an excellent account of their results and of the more general approach of Jacques (1969), see White (1973a), as well as Ringel (1974).

Theorem 2-26 For $n \geq 1$,

$$\gamma(K_n) = \left\{ \frac{(n-3)(n-4)}{12} \right\}$$

PROOF Again, half the proof is simply Corollary 2-8. The other half involved highly subtle and delicate combinatorial techniques, and was almost as difficult a problem as the 4CC itself.

Recently, Gross and Alpert (1973a, 1973b) and Gross and Tucker (1974) have employed more topological methods, such as the theory of branched coverings, to reobtain and extend the Ringel–Youngs constructions.

In Fig. 2-29, we give an example of a branch covering of a trivial map by a toroidal map M. The trivial map consists of a single vertex v and loop e which divides the sphere into two regions, one of which is shaded and lies to the right of the loop (i.e., inside it) which has been assigned the sense of direction indicated by an arrowhead. The map M is drawn on a rectangle whose sides are identified to create a torus; and each vertex is mapped to the vertex v below. Each edge of M,

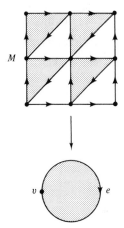

Figure 2-29

assigned a direction as indicated, is mapped to e and the mapping is extended to each region of M so that shaded regions map to the interior of the loop.

For a discussion of other interesting problems arising in connection with graphs and surfaces, see the survey paper by Kainen (1974b).

2-5 SOME COLORING ALGORITHMS

Given a reasonably large graph, say with 20 vertices or more, it is extremely difficult to see how to color it with much efficiency. To this end, we present a very simple and elegant result of Welsh and Powell (1967). Basically, the idea of the algorithm is to be greedy. Pick a vertex and color it red. Proceed to another vertex not adjacent to anything red and color it red. When all vertices which can be colored red are exhausted, color some vertex blue and go through all the vertices again. Keep introducing new colors until all vertices are colored. The trick, and it is a very reasonable one, is to first order the vertices in terms of nonincreasing degrees so that the first vertices considered are those of highest degree which might be expected to cause maximum problems.

Now let us give a precise statement of their method.

Algorithm Let G be a graph with vertices $v_1, \ldots, v_n$ listed in such a way that the corresponding degrees are nonincreasing $d_1 \geq d_2 \geq \ldots \geq d_n$. Let $T_1 \subset V(G)$ be defined as follows.

Put $v_1 \in T_1$. If $v_{i_1}, \ldots, v_{i_m} \in T_1$ $(1 = i_1 < \cdots < i_m)$, put $v_j \in T_1$ if $j > i_m$ and v_j is not adjacent to any vertex already in T_1. To define T_2, let i be the least integer such that $v_i \notin T_1$. Put $v_i \in T_2$. If $v_{j_1}, \ldots, v_{j_p} \in T_2$ $(2 \leq j_1 < \cdots < j_p)$, put $v_m \in T_2$ if $m > j_p$ and v_m is not adjacent to a vertex already in T_2. Continuing in this way, define a sequence T_i of disjoint subsets of $V(G)$ such that $V(G) = \bigcup_{i=1}^{\infty} T_i$. Of course, for r sufficiently large, $T_r = \emptyset$ and we set $\alpha(G)$ equal to the largest index i for which $T_i \neq \emptyset$. Obviously, G can be $\alpha(G)$-colored.

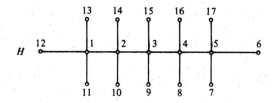

Figure 2-30

Theorem 2-27 $\alpha(G) \le k(G) = \min \{j \mid j \le d_j + 1\}$.

PROOF   Before proving the theorem, we first note that $k(G) = \max_i$ $\min \{d_i + 1, i\}$. Now, by construction, every vertex v_i in $V(G) - \bigcup\limits_{j=1}^{m} T_j$ is adjacent to at least one vertex in each of the sets $T_1, \ldots, T_m$. Hence, $d_i \ge m$. Therefore, $v_i \in \bigcup\limits_{j=1}^{d_i+1} T_j$. But, by definition, $v_i \in \bigcup\limits_{j=1}^{i} T_j$ and so $v_i \in \bigcup\limits_{j=1}^{\phi(i)} T_j$, where $\phi(i) = \min \{i, d_i + 1\}$. Hence, $V(G) \subset \bigcup\limits_{j=1}^{k(G)} T_j$ since $k(G) = \max_i \phi(i)$; so $\alpha(G) \le k(G)$.

Consider the following graph H (Fig. 2-30) labeled so that $d_1 \ge \cdots \ge d_{17}$.

For this tree, $k(H) = 5$, which is not too impressive. On the other hand, $\alpha(H) = 2$. Note that α actually depends upon the ordering of the vertices. For example, if we reorder the vertices of H as in Fig. 2-31, we get $\alpha(H) = 3$; $T_1 = \{1, 2, 6, 7, 9, 10, 14, 15, 17\}$, $T_2 = \{3, 4, 8, 11, 12, 13, 16\}$, $T_3 = \{5\}$.

The reason for this ambiguity in $\alpha(H)$ is that H has many vertices of the same degree. Heuristic techniques which utilized orderings based on the degrees of vertices adjacent to a given vertex have shown some potential for improving these results (see Williams, 1970).

We saw that, while the theoretical upper bound remained the same, the number of colors actually obtained by applying the algorithm depends on the particular ordering of the vertices. Williams has suggested an interesting heuristic modification. Order the n vertices of a graph G in any fashion at all, say $v_1, \ldots, v_n$. Define a column vector d^0 of length n by $d_i^{\,0} = 1$ $(1 \le i \le n)$. Inductively, for $j \ge 0$, define $d^{j+1} = Ad^j$, where $A = (a_{rs})$ is the $n \times n$ matrix whose (r, s) entry a_{rs} is equal

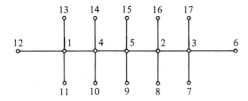

Figure 2-31

to 1 if v_r is adjacent to $(\sim)v_s$, and to 0 otherwise. Note that d^1 is just the vector of degrees—see (a) below. Williams' idea is to use d^k, k large, instead of d^1 to order the vertices before applying the Welsh–Powell method. The vectors d^k converge to the dominant eigenvector of A, and this convergence seems computationally to occur after only about $n^{1/3}$ iterations.

(a) Prove that $d_i^1 = \deg(v_i)$ and that $d_i^2 = \sum_{v_j \sim v_i} \deg(v_j)$.

(b) Use d^2 with the Welsh–Powell method to color the graph in Fig. 2-30.

Very few constructions have been given which show constructively how to color some general class of maps. The following scheme due to Ringel (1959) does show us how to three-color the edges (so that no two edges with a common endpoint are colored the same) of a particular kind of cubic map. Later we will see how to four-color any cubic map whose edges have been three-colored. A map is *bridgeless* if every edge separates two regions.

Let M be a cubic bridgeless map. Suppose that the number of edges in the boundary of every region is a multiple of 3. Call the three colors 1, 2, and 3, and give them the usual cyclic ordering so that 2 follows 1; 3 follows 2; and 1 follows 3. If e, f, and g are the three edges of M incident with some vertex, give them the cyclic ordering induced by the clockwise orientation of the plane; that is, f follows e if, moving clockwise from e, we first encounter f.

Coloring scheme Begin with some edge e of M and color it arbitrarily, say with 1. Now consider the four edges adjacent to e, two at each endpoint. In the cyclic orderings at each endpoint, these four edges either follow or precede e. Give them

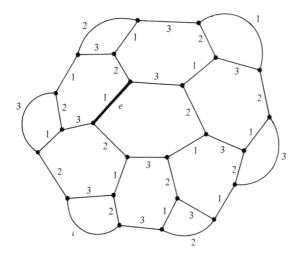

Figure 2-32

the corresponding color. (Thus, if f follows e, color f with 2.) Continue the process until all edges have been colored (see Fig. 2-32).

In order to check that this procedure is unambiguous—in other words, that only one color is assigned to each edge—and that no two adjacent edges receive the same color, it suffices to observe that, since the boundary of every region in M has length $\equiv 0 \pmod 3$, every cycle in M has length $\equiv 0 \pmod 3$ (see page 120).

This provides us with a method for four-coloring the regions of a cubic bridgeless map in which the number of sides of every region is a multiple of 3.

THREE

SOLUTION OF THE FOUR-COLOR PROBLEM; CUBIC MAPS, REDUCIBILITY, AND DISCHARGING

3-1 INTRODUCTION

In this chapter, we shall describe the current solution of the four-color theorem. Section 3-2 develops the theory of triangulations and shows how any planar graph may be "standardized" in such a form. In Sec. 3-3 we continue with the early results involving reducibility. The elementary properties of reducible configurations are derived and a list of important reducible configurations is presented.

Section 3-4 treats discharging and the concept of unavoidable sets. Then *D*-reducibility and the use of the computer are studied in the next section. This completes our description of the Appel–Haken proof.

In Sec. 3-6, we explain the probabilistic and heuristic techniques which led to the proof. This methodology is, in our opinion, one of the most important aspects of the proposed proof. We discuss our conclusions on philosophical ramifications in the final section.

3-2 STANDARDIZING GRAPHS AND MAPS— TRIANGULATIONS

Maps occur in a variety of forms. Many regions can meet at a single point, or one region can completely surround another. For example, a map may look like Fig. 3-1.

We will show that we need only consider maps of a certain "standard" kind in our discussion of the 4CC. In order to do this, we first go through a corresponding

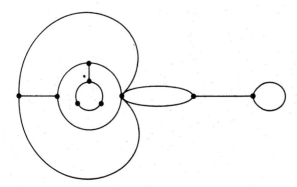

Figure 3-1

standardization of the problem for graphs, making the 4CC correspond to coloring the vertices of a maximal planar graph. Then by carefully investigating the properties of a maximal planar graph and of any map whose dual is such a graph, we obtain standard maps.

Suppose that M is a map and $G = D(M)$. An edge e in M which is a bridge lies on the boundary of only one region and so, by the definition of $D(M)$, e gives rise to a loop e' in $D(M)$. Let R and S be regions of M. We say that R and S are *multiply adjacent* (and that M has *multiple adjacencies*) if R and S contain more than one common edge in their boundaries. By the definition of $D(M)$, $D(M)$ has parallel edges if and only if M has multiple adjacencies (see Fig. 3-2).

We will use the *shrinking* operation to eliminate certain edges. Suppose e is an edge of M. Let M/e denote the map obtained from M by shrinking e to a point. Then $U(M/e) = U(M)/e$ and $D(M/e) = D(M) - e^*$, where e^* is the edge of $D(M)$ dual to e. Thus, by a sequence of edge-shrinkings, we can get rid of all bridges and multiple adjacencies in M; this process corresponds precisely to the removal of all loops and all parallel edges from $D(M)$ (see Fig. 3-3).

Now if G is any graph and we are interested in four-coloring (or k-coloring) G, it is obvious that neither loops nor parallel edges make any difference. Thus, to color a graph, it suffices to color the simple graph obtained from it by deleting loops and parallel edges. Hence, to color a map M, we need only color

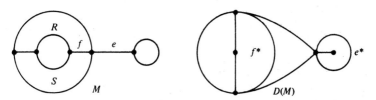

Figure 3-2

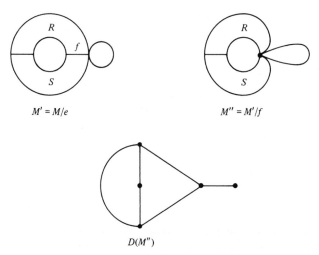

$M' = M/e$ $M'' = M'/f$

$D(M'')$

Figure 3-3

the map obtained from M by shrinking edges to get rid of bridges and multiple adjacencies.

Maximal Planar Graphs

Definition A *maximal planar graph* is a planar graph to which no new edges can be added without violating planarity.

Lemma 3-1 If G is a planar graph in which some region R has on its boundary four or more vertices, some pair of these vertices cannot be adjacent.

PROOF If all pairs of vertices of R are adjacent (with edges drawn exterior to R), we can add a vertex in the interior of R and join it (without edge-crossings) to each vertex without violating planarity. But, in fact, the result for the vertices of R plus the new vertex is a complete graph on five or more vertices which we know to be nonplanar. Hence, some pair of vertices on the boundary of R cannot be adjacent.

Theorem 3-1 Let G be a planar graph. G is maximal planar if and only if every region including the unbounded region has three sides.

PROOF Let G be maximal planar and assume that some region R of G has four or more vertices on its boundary. By the lemma, some pair of these vertices is nonadjacent. We can join them by an edge in G and the resulting graph is still planar. This contradicts the fact that G was maximal planar.

Conversely, if every region of a planar graph G is bounded by three sides, that is, $3r = 2m$ or $m = 3n - 6$, G must be maximal planar, i.e., we cannot add any more edges to it since it has the maximum number of edges a planar graph can have.

Definition A *triangulation* is a planar graph in which every region is bounded by three sides.

Triangulations have many pleasant properties not shared by more general planar graphs. One such property which we shall need later (in the proof of Theorem 3-7) involves the important notion of k-*connectedness*.

A connected graph G is (at least) k-*connected* if there is no set of $k - 1$ vertices whose removal makes G disconnected or trivial. Note that a three-connected graph could be four-connected as well. A set S of vertices whose removal disconnects G, and with $|S| = k - 1$, is a *minimal* disconnecting set for a k-connected graph G. For G a triangulation, we show in the Appendix that for a minimal disconnecting set S the subgraph $G(S)$ of G induced by S must consist of a circuit. This immediately implies that any triangulation must be three-connected.

From the theorem, triangulation is equivalent to maximal planarity and, hence, we shall use it as a more descriptive geometric term for maximal planar graphs.

Now let G be any simple planar graph. Then G is a subgraph of a maximal planar graph $\overline{G}$, obtained from G by adding edges, and four-coloring $\overline{G}$ thereby four-colors G.

Any maximal planar graph which is obtained from a planar graph G by first deleting loops and parallel edges and then adding new edges will be called a *standardization* of G, and the process will be called *standardizing* a graph.

We summarize our results in the following theorem.

Theorem 3-2 To prove Conjecture C_1, it suffices to be able to four-color every maximal planar graph. That is, we may standardize our graph first before trying to four-color it.

Maximal planar graphs have several nice properties. For example, every maximal planar graph G is three-connected. As an immediate corollary, every maximal planar graph is bridgeless (all we need is that the graph be two-connected), for if e is a bridge, then removing either endpoint of e also removes e and, hence, disconnects the graph. Moreover, if $G \neq K_3$ is drawn in the plane to produce a triangular map T, then any two regions must intersect either at a vertex or along an edge (with both endpoints), if they intersect at all. They cannot intersect along two edges for this would force G to have a pair of parallel edges. In particular, then, no two regions in any drawing of G are multiply adjacent.

Let $\mathcal{M}$ be any collection of planar maps. We say that a collection $\mathcal{G}$ of planar graphs is *dual* to $\mathcal{M}$ if

$$\mathcal{G} = \{G \,|\, G = D(M) \text{ for some } M \in \mathcal{M}\}$$

Obviously, if $\mathcal{G}$ is dual to $\mathcal{M}$ and we can four-color every graph in $\mathcal{G}$, then we can four-color every map in $\mathcal{M}$.

This suggests that we now explore the following question. If $\mathcal{G}$ is the class of maximal planar graphs, then to what class $\mathcal{M}$ of maps is $\mathcal{G}$ dual? In order for this sort of question to be meaningful, we ought to know that every planar graph is the dual of some map. (Previously, all we showed was that if G is planar, then there exists a map M such that $G = D(M)$.) To this end, we prove the next result, due to Whitney (1933c). The proof of this theorem should be read after Sec. A-2 of the Appendix. A mapping that is one-to-one and onto is known as a *bijection*.

Theorem 3-3 Let G be any connected planar graph and let M be any map with underlying graph G. Then G is isomorphic to $D(M^*)$.

PROOF Let $G^* = U(M^*)$ and $G^{**} = U(M^{**})$. Then G^* is a Whitney dual to G and G^{**} to G^*. Let $W: E(G) \to E(G^*)$ and $W^*: E(G^*) \to E(G^{**})$ be the corresponding bijections. Then $W^{**} = W^*W: E(G) \to E(G^{**})$ is also a bijection and W^{**} carries $Z(G)$ isomorphically onto $Z(G^{**})$ and $Z^*(G)$ onto $Z^*(G^{**})$. Let $e^{**} = W^{**}(e)$ and, if $v \in V(G)$, let v^{**} denote the corresponding vertex in $V(G^{**})$. We need to prove that if $e = [v, w]$, then $e^{**} = [v^{**}, w^{**}]$. But this is clear since, if $e = [v, w]$, then $W(e)$ lies on the common boundary of the two regions of M^* corresponding to v and w and hence $W^*(W(e)) = e^{**}$ joins v^{**} and w^{**}. Thus, $e \to e^{**}$ is an isomorphism of graphs; that is, $G \cong D(M^*)$.

We really needed to use the geometric derivation of the bijections W and W^* in this proof since otherwise we would be proving that any two Whitney duals to the same graph are isomorphic—a fact we know to be false.

Note that if $Z_1, \ldots, Z_s$ is the cycle basis of $Z(U(M))$ corresponding to the boundaries of regions in M and Z_0 is the boundary of the remaining region, then the corresponding cycles $Z_0^{**}, Z_1^{**}, \ldots, Z_s^{**}$ in $Z(U(M^{**}))$ satisfy the hypothesis of MacLane's theorem for planarity (see Appendix, Theorem A-3) and, hence, are the boundaries of the regions of M^{**}. In other words, M and M^{**} are isomorphic as maps. (We have actually shown that the ismorphism $U(M) \cong U(M^{**})$ carries the boundary of a region to the boundary of a region.) Now let us return to the original question.

Theorem 3-4 Let $\mathcal{G}$ be the class of maximal planar graphs. Then $\mathcal{G}$ is dual to the class $\mathcal{M}$ of all maps M satisfying:

(a) M is cubic.

(b) $U(M)$ is simple.

(c) M is bridgeless and has no multiple adjacencies between regions.

PROOF The following statements are obvious:

(1) M is cubic if and only if $D(M)$ is maximal planar.

(2) M has no bridges or multiple adjacencies if and only if $D(M)$ is simple.

Dualizing (2) we get

(3) $U(M)$ is simple if and only if M^* has no bridges or multiple adjacencies.

Now, (a) follows directly from (1) and (c) from (2). To get (b), we just use (3) and the observation, made after Theorem 3-2, that the map formed by drawing a maximal planar graph has no bridges or multiple adjacencies.

Let us use this result to dualize Theorem 3-2.

Theorem 3-5 To prove Conjecture C_0, it suffices to four-color every map $M \in \mathcal{M}$.

PROOF Let M be any map and put $G = D(M)$. Let G' be a standardization of G. By the preceding theorem, $G' = D(M')$ for some map $M' \in \mathcal{M}$. Now four-coloring M' is equivalent to four-coloring G'. But any four-coloring of G' induces a four-coloring of G and, hence, of M.

Theorems 3-2 and 3-4 now show the equivalence of the two following conjectures to Conjectures C_0 and C_1, respectively.

Conjecture C_2 Let $M \in \mathcal{M}$. Then M can be four-colored.

Conjecture C_3 Let $G \in \mathcal{G}$. Then G can be four-colored.

For completeness, let us state one more conjecture.

Conjecture C'_2 Let M be any cubic bridgeless map. Then M can be four-colored.

Clearly, C'_2 implies C_2. Conversely, C_2 implies C_0 implies C'_2.

The proof of Theorem 3-2 shows us how to take any map M and replace it with a *standardization* $M' \in \mathcal{M}$; that is, it shows us how to *standardize* any map. As we have seen, removing all loops in $D(M)$ corresponds to shrinking all bridges in M to points, and removing parallel edges in $D(M)$ corresponds to shrinking to points all edges save one on the common boundary of two regions in M. Finally, adding edges to the resulting simple graph $D(M)$ corresponds to "splitting" vertices in M, as indicated below in Fig. 3-4.

Thus, starting with any map M and using the operations of edge-shrinking and vertex-splitting, we can produce a standardized map M' in $\mathcal{M}$ such that four-coloring M' induces a four-coloring of M.

There is another, more traditional approach that we should mention here. Let M be any map. If M contains any bridges, shrink them to points to obtain a new map M_1. Now consider $M_2 = T(M_1)$, the cubic map obtained by "inflating"

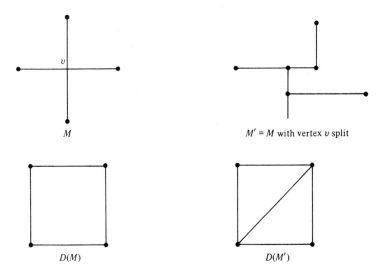

M $M' = M$ with vertex v split

$D(M)$ $D(M')$

Figure 3-4

every vertex of M (see Fig. 1-12). If $T(M_1)$ has a bridge e, then e must correspond to some edge f in M_1 and f is a bridge. But M_1 is bridgeless, so M_2 is bridgeless.

Since M_2 is cubic and bridgeless, it is also loopless. M_2 may, however, have parallel edges and multiple adjacencies. The latter can be gotten rid of back at the M_1 stage, along with the bridges, and will not reappear in M_2. If M_2 has parallel edges, then we can take $M_3 = T(M_2)$. M_3 is now cubic and bridgeless with $U(M)$ and $D(M)$ simple; that is, $M_3 \in \mathcal{M}$. Clearly, a four-coloring of M_3 induces a four-coloring of M.

Searching for a Minimal Counter-Example

Minimal graphs Suppose that the 4CC is false. Then there is some planar graph G with $\chi(G) = 5$. Among all such five-chromatic planar graphs one with a minimum number of vertices is called a *minimal* graph. Thus, a planar graph G is minimal if $\chi(G) = 5$ but $\chi(H) \le 4$ whenever H is planar and has fewer vertices than G.

In seeking a proof of the 4CC, one should expect minimal graphs to be exceedingly rare. Thus, by deriving enough information about minimal graphs, one is expected to prove their nonexistence.

Let G be a minimal graph. We claim that G must, in fact, be a triangulation. Suppose that G is embedded in the plane so that some region has more than three sides. Choose two vertices v and w in the boundary of the region which are not themselves adjacent. (The reader should verify that this is possible.) Add a new edge $[v, w]$ and now contract it. The resulting graph G can be four-colored since it has fewer vertices than G. But this induces a four-coloring of G (in which v and w receive the same color). Hence we have the following theorems.

Theorem 3-6 If G is minimal, then G is maximal planar.

Theorem 3-7 If G is minimal, then G is five-connected.

PROOF Let S be a minimal disconnecting set in G. We must show that $|S| \geq 5$. From the corollary concerning a minimal disconnecting set for a maximal planar graph (see Appendix), we known that $G(S)$ is a circuit. If $|S| = 3$, $G(S)$ is a triangle. Draw G in the plane. Let A denote that portion of G interior to $G(S)$, together with $G(S)$, and let B denote the corresponding external portion of G also including $G(S)$. By the minimality of G, A and B can be four-colored: say a is a four-coloring of A and b of B, where both a and b use the same set of four colors. The three vertices v_1, v_2, v_3 in S receive colors $a(v_1), a(v_2), a(v_3)$ and $b(v_1), b(v_2), b(v_3)$, respectively. Now, by permuting the colors in b, if necessary, we can insure that $a(v_i) = b(v_i)$, $i = 1, 2, 3$. This yields a four-coloring of G which is impossible.

Suppose $|S| = 4$. Then $G(S)$ is a quadrilateral with vertices x, y, w, z (listed in clockwise order). Let A and B have the same meaning as above.

Since $A + [y, z]$ is planar and has fewer vertices than G, A has a four-coloring a_1 such that $a_1(y) \neq a_1(z)$. Similarly, B has a four-coloring b_1 such that $b_1(y) \neq b_1(z)$. Relabeling if necessary, let $\alpha = a_1(y) = b_1(y)$, $\beta = a_1(z)$, and $\gamma = a_1(x) = b_1(x)$. If $a_1(w) = b_1(w)$, we are done. Hence, suppose $a_1(w) = \gamma$ but $b_1(w) = \delta$. If there is no γ–δ path in B from x to w, then we may use a Kempe color-chain argument (as in Chap. 2) to change the coloring b_1 into a coloring b_2 with $b_2(w) = \gamma$ so that a_1 and b_2 agree on S. Suppose that there is a γ–δ path from x to w in B. Then B has no α–β path from y to z, so we may change b_1 into a coloring b_3 with $b_3(y) = \alpha = b_3(z)$, $b_3(x) = \gamma$, and $b_3(w) = \delta$. Now return to A. Since $A + [x, w]$ is also planar, we can find a four-coloring a_2 of A with $a_2(x) = \gamma$, $a_2(w) = \delta$, and $a_2(y) = \alpha$. If $a_2(z) = \alpha$, a_2 matches b_3 on S. If $a_2(z) = \beta$, then a_2 matches b_1 on S. This completes the proof.

Let G be a minimal graph. G must contain a vertex v of degree 5, and we write G_v for the complement $G-v$ of v in G. If G_v is embedded in the plane, the boundary of every region is a triangle with the exception of one pentagon (called the *boundary* of G_v) whose vertices are the neighbors of v in G. By the minimality of G, G_v is four-chromatic but, in any four-coloring c of G_v, all four colors must appear in the boundary of G_v; otherwise, c would extend to a four-coloring of G. Following Tutte (1972c) we call G_v a *chromatic obstacle*. Of course, the existence of a chromatic obstacle is equivalent to the falsity of the 4CC.

Minimal maps These are defined in the obvious way and have the appropriate (dual) properties to those of minimal graphs. For example, the notion of "separating circuit" in a graph G becomes *ring* in the map M to which it is dual. If S is a minimal separating set in G, then the ring corresponding to $G(S)$ is called *proper*, and Theorem 3-7 yields the fact that a minimal map has no proper k-ring for $k < 5$. (See Sec. 3-3 for these notions.)

Moreover, a minimal map M has no countries with fewer than five sides, and is bridgeless and cubic. In fact, $U(M)$ is three-connected since it is the dual of the minimal graph $G = D(M)$, and by a result of Whitney (1932a) the dual of a three-connected graph is three-connected.

3-3 REDUCIBILITY AND FOUR-COLORING

The idea of reducibility in maps and graphs (particularly in minimal graphs) which played a central role in solving the 4CC essentially takes out a subgraph from a graph G, four-colors the resulting graph (or modifications of it), puts the subgraph back, and four-colors G. If all planar graphs contained subgraphs which offer such flexibility, the 4CC would be true.

We have already encountered simple versions of reducibility by taking out a vertex, coloring the resulting graph, then putting the vertex back, and obtaining a proper coloring of the entire graph.

Removing vertices of degree ≤ 3 is easy. We do not even need planarity. Simply four-color the remainder of the graph and assign one of the four colors unused by its ≤ 3 neighbors to the vertex. For a degree 4 vertex v in a planar graph, we must be a little more careful. If the four neighbors do not all get distinct colors as before, we can use one of the remaining colors for v.

If the four neighbors (in clockwise order w_1, w_2, w_3, w_4) do get distinct colors—say red, green, blue, and yellow, respectively—then we can use an argument exactly like our first proof of the five-color theorem. Try to change w_1 from red to blue. This is possible unless there is an alternatingly red and blue chain of vertices (called a red–blue Kempe chain) joining w_1 and w_3 in $G-v$.

In this case, try to change w_4 from yellow to green without changing w_2 from green to yellow. This is possible unless there is a green–yellow Kempe chain from w_2 to w_4 in $G-v$. By the planarity of G, both of these Kempe chains cannot exist simultaneously. Hence, one way or the other, we can change the four-coloring of $G-v$ so that it extends to G.

We have now seen that any vertex v of degree ≤ 4 is reducible in that the four-colorability of a planar graph can be reduced to the four-colorability of the graph with the vertex removed.

We have also seen that every planar graph has a vertex of degree ≤ 5. Thus, we only need to concern ourselves with planar graphs whose minimum degree vertex is 5, for otherwise we could four-color inductively. In general, a vertex of degree 5 is not reducible. This is the essence of Heawood's counter-example to Kempe's attempted proof of the 4CC. Later we shall see that certain configurations of degree 5 vertices (and of higher degree vertices as well) are reducible.

The idea of reducibility now is to remove subgraphs from a planar graph so that the four-colorability of what is left implies the four-colorability of the original graph. One problem, however, is that the subgraph may sit in its graph in many different ways. To simplify matters, and without loss of generality, we shall only consider planar graphs in which each region has three sides. Such a graph may be obtained by joining pairs of vertices on the boundary of each region until no more edges can be drawn without crossing. If we can four-color such triangulations, we can also four-color the planar graph which gives rise to it. These triangulations restrict the number of ways in which a subgraph may be contained. For example, a vertex of degree 5 may sit in a planar graph in many different ways. But if it is in a triangulation, its five neighbors must induce a circuit so that it looks as in Fig. 3-5.

A *configuration* consists of a subgraph with specification of its vertex degrees and the manner in which it is embedded in a planar triangulation. A configuration R is *reducible* if the four-colorability of any planar graph G which contains R can be deduced from the four-colorability of planar graphs G with fewer vertices. The graph G' with fewer vertices than G from which the four-colorability of G can be deduced need not be a subgraph of G. Let us illustrate this with an example.

A circuit Q in a planar graph G is a *separating* circuit if removing Q (vertices and edges both) from G produces two (nonempty) components G_1 and G_2. Q is nonseparating if and only if it is the boundary of a region.

If a planar graph G has a *nonseparating* circuit of length > 3, then the region bounded by this circuit has a pair of nonadjacent vertices in its boundary (see Lemma 3-1). Identifying these vertices produces a smaller graph G' which is planar. Moreover, any four-coloring of G' induces a four-coloring of G, so G is reducible. Thus, a nonseparating circuit of length > 3 is a reducible configuration.

This argument is exactly parallel to that which proves Theorem 3-6. Call a (planar) graph *irreducible* if it contains no reducible configurations. We have just shown that an irreducible graph is maximal planar. Since a minimal graph is certainly irreducible, our result here actually generalizes Theorem 3-6.

To show that the four-colorability of a planar graph G depends on the four-colorability of a smaller graph G', we need only find a reducible configuration in G. Given a configuration R and an arbitrary planar graph G containing it, if G contains another known-reducible configuration, then the four-colorability of G certainly depends on the four-colorability of some smaller graph G'. Thus, to prove R reducible, we may suppose that its containing graph G has no other

Figure 3-5

reducible configurations. *According to what we have already proved, this means we may assume that G is a triangulation and has no vertex of degree smaller than* 5.

There is one more important property of G we would like to be able to mention; this is the reducibility of a separating circuit of length < 5. This could be proved exactly as in Theorem 3-7. Let us summarize our results so far in a theorem.

Theorem 3-8 The following configurations in a planar graph are reducible: (*a*) vertex of degree less than 5; (*b*) region with boundary length greater than 3; (*c*) separating circuit of length less than 5.

This allows us to assume that the graph in which any configuration is embedded is a triangulation and has minimum degree ≥ 5. Furthermore, (*c*) will tell us something about the ways in which configurations can be embedded. Such graphs are called *suitable*.

Given a configuration embedded in the above kind of planar graph, its boundary vertices may come equipped with specified degrees. For example, consider the configuration B consisting of a vertex of degree 5 (v_5 for short) with three consecutive five-neighbors. We represent B in Fig. 3-6 below in which, following Heesch, the symbol "." is used to denote a vertex of degree 5.

Suppose B is embedded in G. Since G is a triangulation with no separating three-circuit, the embedding must include two additional lines to produce the configuration in which each of the triangles are the boundaries of regions (Fig. 3-7).

In Figure 3-7 we have shown the extended configuration (also called B). The dashed lines (called *legs*) represent the other edges which must be present because of the degree constraints on the boundary vertices. Since G is a triangulation, the two legs emanating from the endpoints of an edge in B must themselves have a common endpoint not in B (see Fig. 3-8). The symbol "ʊ" denotes a vertex of unspecified degree.

Furthermore, the endpoints of any two other legs, which do not form a triangle with some edge in B, must be distinct or G would contain a separating three-circuit. Adding a few additional edges because of the triangularity of G and redrawing B in a slightly more symmetric fashion, we obtain a complete picture of all vertices adjacent to B (the "first neighborhood" of B) (see Fig. 3-9).

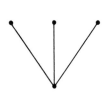

Figure 3-6

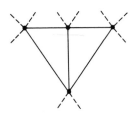

Figure 3-7

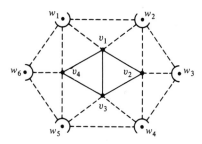

Figure 3-8

Figure 3-9

One way to show that this configuration is reducible might be to show that any four-coloring of the vertices of the boundary hexagon (ring of six vertices) must extend to the interior. But this sort of argument is more than we need to show. If we can change the coloring in some way to force it to have certain specified properties, then we may be able to extend it over the interior. After all, we only need the existence of *some* coloring.

This is exactly what we shall do. Imagine that our configuration appears in some triangulation G. We create a new graph G' by removing the four vertices of our configuration, leaving in the boundary vertices $w_1, \ldots, w_6$ which now bound a hexagonal region. We continue to modify G' by adding edges between w_2 and w_6 and between w_4 and w_6, and by identifying the vertex w_2 with w_4. The exterior of the hexagon has not changed and our modifications have not destroyed planarity. Let us call the new graph $\overline{G}$; Fig. 3-10 reminds us of what was done to the hexagon. A single line means an edge. A double line indicates identification of the vertices.

The point of this procedure is that any four-coloring of $\overline{G}$ must assign the same color to w_2 and w_4 (since they are the same vertex in $\overline{G}$) and a different color to w_6. Thus, any four-coloring of $\overline{G}$ must assign a coloring of the type (r, b, x, b, y, g) to the six vertices $(w_1, \ldots, w_6)$, where x and y denote unspecified colors assigned to w_3 and w_5. It turns out that all such colorings are extendable

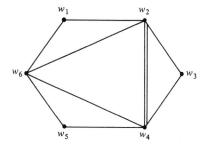

Figure 3-10

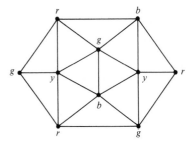

Figure 3-11

to the four interior vertices v_1, v_2, v_3, v_4 with one exception: the coloring (r, b, r, b, r, g) does *not* extend, since v_1 and v_3 must be colored (y, g) or (g, y), and hence v_2 is forced to be adjacent to vertices of all four colors. In this latter case, however, a Kempe-chain argument can be made as follows. Either w_1, w_3, and w_5 all belong to the same r–y component (in which case we could change the color of w_4 from b to g and extend the coloring to all of G, as shown in Fig. 3-11) or else w_1, w_3, w_5 do not all belong to the same r–y component and so one of the three vertices belongs to a r–y component not containing the other two vertices. If this vertex is w_3, we can change the colors of w_1 and w_5; if the isolated vertex is w_5 (or w_1), we can change its color. Each of the modified colorings can be extended (see Fig. 3-12).

Of course, more elaborate configurations may require much more complex arguments and, as we have seen, not all configurations are reducible. Nevertheless, the hope of reducibility theory has been to find a sufficiently rich list of reducible configurations that every planar triangulation must contain one.

This wish has apparently now been fulfilled. Haken and Appel, using the methodology of Heesch, have a list of reducible configurations of 1936, at least one of which must occur in any planar triangulation. Their story and the technique of discharging by which they found the list of configurations and the computer-assisted reducibility arguments is given in the next three sections.

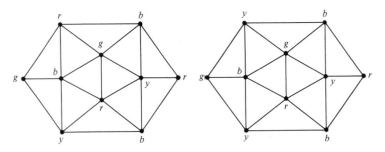

Figure 3-12

It will now be convenient to formalize what we have done. Given a vertex v in a triangulation G, let us define its *first neighborhood* (or *star*) to consist of the subgraph of G induced by v and all other vertices of G adjacent to v. The boundary of the first neighborhood is called the *link* of v and is produced by removing v from its star. Similarly, we can define star and link for other configurations as well. For example, the star of the first neighborhood of v is called the *second neighborhood*.

We can now generalize our result that, whenever the configuration B is embedded in a triangulation, the embedding is extendable as indicated in Fig. 3-9.

Given any configuration R of diameter ≤ 2, the first neighborhood of R in G is reconstructable from the degree constraints on boundary vertices using only the fact that G does not contain any of the three reducible configurations of Theorem 3-8. This will be the justification for drawing the first neighborhood (or a portion of it) for some configurations.

Now we already know that a suitable graph G cannot contain any vertices of degree less than 5. Moreover, since G is five-connected, it cannot contain any separating circuit of length less than 5.

Theorem 3-9 (Birkhoff, 1913) Let Z be a separating circuit of length 5 in a suitable graph G and let A and B denote the two components of $G-Z$. Then either A or B is trivial. (In other words, a five-circuit which separates two nontrivial components is reducible.)

PROOF Assume that neither A nor B is trivial. Then both G/A and G/B are planar graphs with fewer vertices than G and so have four-colorings c_A and c_B, respectively. But a four-coloring of G/A (G/B) is just a four-coloring of $G-A$ ($G-B$) which induces a three-coloring of Z. Let c' and c'' denote the 2 three-colorings of Z induced by c_A and c_B, respectively. Now, up to permutation of colors, a three-coloring d of Z is uniquely determined by specifying which vertex v of Z is colored differently from the remaining four vertices. We call v the *marked vertex of Z* with respect to the coloring d and write $v = m(d)$.

If $m(c') = m(c'')$, then it is plain to see that the colorings c_A and c_B induce a four-coloring c of G. Now suppose $m(c') \neq m(c'')$. Enumerate the vertices of Z as $v_1, \ldots, v_5$ and assume, without loss of generality, that $m(c') = v_1$. Either (1) $m(c'')$ is adjacent in Z to v_1, or (2) it is not.

We claim that case (2) can be reduced to case (1). That is, if case (2) holds, c_A and c_B can be changed to obtain colorings d_A and d_B for which the corresponding marked vertices $m(d')$ and $m(d'')$ are adjacent in Z. This is done as follows.

Again, without loss of generality, let $m(c'') = v_3$. Then $c' = (\gamma, \alpha, \beta, \alpha, \beta)$ (meaning, of course, $c'(v_1) = \gamma$, $c'(v_2) = \alpha = c'(v_4)$, $c'(v_3) = \beta = c'(v_5)$) and $c'' = (\alpha, \beta, \gamma, \alpha, \beta)$. If there is no β–γ path in $G-A$ from v_2 (and v_3) to v_5, then a Kempe-chain argument would allow one to change c_B into d_B so that $d'' = (\alpha, \gamma, \beta, \alpha, \beta)$. But $m(d'') = v_2$, which is adjacent to v_1.

If there is a β–γ path in G–A from v_2 to v_5, then there is no α–δ path in G–A from v_1 to v_4 and so we can change c_B into d_B such that $d'' = (\alpha, \beta, \gamma, \delta, \beta)$. Now let d_A be any four-coloring of G–B in which v_2 and v_5 obtain the same color. Then $d' = (x, \alpha, y, z, \alpha)$, where x, y, z denote three not-necessarily-distinct colors. Without loss of generality, let $y = \beta$ and $z = \gamma$. If $x = \beta$, then $m(d') = v_4$, which is adjacent to $v_3 = m(c'')$. If $x = \gamma$, $m(d') = v_3 = m(c'')$, which is the easy case. Finally, if $x = \delta$, then, up to a permutation of the colors, $d' = d''$.

Now we consider case (1) and assume that $m(c'') = v_2$. So $c' = (\gamma, \alpha, \beta, \alpha, \beta)$ and $c'' = (\beta, \gamma, \alpha, \beta, \alpha)$. If there is no β–γ path in G–A from v_2 to v_4, change the color at v_4 to γ so that $m(c'') = m(c')$. If there is a β–γ path in G–A from v_2 to v_4, then there is no α–δ path between v_3 and v_5 and we may change c_B into d_B so that $d'' = (\beta, \gamma, \delta, \beta, \alpha)$. Now four-color G–B with some coloring d_A in which v_1 and v_4 get the same color; so $d' = (\beta, x, y, \beta, z)$. If $x = \gamma$ and $y = \delta$, then we may have $z = \alpha$, in which case $d' = d''$. Or we may have $z = \delta$, in which case $d' = (\beta, \gamma, \delta, \beta, \delta)$ so $m(d') = v_2 - m(c'')$.

The remaining possibility is $z = \gamma$, that is, $d' = (\beta, \gamma, \delta, \beta, \gamma)$. Hence, $m(d') = v_3$. Thus, from a pair (c_A, c_B) of colorings of G–B and G–A, respectively, with $m(c') = v_1$ and $m(c'') = v_2$, we have obtained a pair (d_A, c_B) with marked vertices $m(d') = v_3$ and $m(c'') = v_2$. Repeating the process once more (interchanging the roles of G–A and G–B), we obtain a pair of colorings (d_A, d_B) with marked vertices $m(d') = v_3$ and $m(d'') = v_4$. Iterating the process twice more yields a pair of colorings (e_A, e_B) with $m(e') = v_5$ and $m(e'') = v_1$. Hence, e_A and e_B induce a four-coloring of G. This completes the proof.

Other reducibility arguments are even worse! Accordingly, we intend to spare the reader the gruesome details. Instead, we shall merely state some of the major results.

First, some terminology. A *k-vertex* is a vertex of degree k. A *k-neighbor* of v is a k-vertex w which is adjacent to v. A vertex v is *minor* if $\deg v \le 6$ and *major* if $\deg v \ge 7$.

Theorem 3-10 (Birkhoff, 1913) A five-vertex with three consecutive five-neighbors is reducible.

Theorem 3-11 (Franklin, 1922) A six-vertex with three consecutive five-neighbors is reducible.

Using additional results due to Franklin (1938), Winn (1937), and Choinacki (1942), one obtains the following theorem.

Theorem 3-12 A five-vertex is reducible if all of its neighbors are minor.

A similar result for six-vertices is established by the work of Bernhart (1947, 1948) and Winn (1937).

Theorem 3-13 A six-vertex is reducible if all of its neighbors are minor.

For seven vertices only results like the following are known.

Theorem 3-14 (Winn, 1938) A seven-vertex with four consecutive five-neighbors is reducible.

Theorem 3-15 A seven-vertex with 5 five-neighbors is reducible unless its other two neighbors are major vertices.

An interesting class of reducible configurations was discussed by Errera (1925). A separating circuit is *chordless* if two nonconsecutive vertices are never joined by an edge (or *chord*) in the containing graph.

Theorem 3-16 A chordless separating circuit is reducible if it consists of an even-length sequence of five-vertices and one (or two) additional vertices.

Corollary 3-1 Let v be an n-vertex in a suitable graph G. Then v has at most $n - 3$ consecutive five-neighbors for n even and $n - 2$ for n odd.

PROOF OF COROLLARY We need only show that the link $\mathrm{lk}(v)$ of v is a chordless separating circuit. Since G is a triangulation with minimum degree ≥ 5, it is easy to see that the neighbors of v, taken in sequential order, induce a separating circuit Z. If this circuit had a chord e joining two nonconsecutive vertices v_i and v_j, then e would have to lie on the outside of Z and so e would determine a separating triangle with $[v, v_i]$ and $[v, v_j]$. But a suitable graph G has no separating circuits of length < 5.

3-4 DISCHARGING AND UNAVOIDABLE SETS

In this section we introduce the idea of discharging due to Heesch and show how it generates an "unavoidable" set of configurations, one of which must be contained in any triangulation.

To discharge a triangulation we must first charge it. We assign a number to each vertex v, called the charge at v, so that the sum of the charges is 12. (In a moment, it will be proved that one can always charge the vertices in this way.) Think of these numbers as electrical charges which can be moved among the vertices. Starting with an initial charge, we may shunt positive charge from one vertex to another (thereby decreasing the charge at the first and increasing it by like amount at the second) so that the new distribution will still sum to 12. Discharging means moving the charges in such a way that no vertex has a positive charge. But this is quixotic. Since the new distribution still has charges summing to 12, some vertices must get positive charge.

This is precisely the point. Any local procedure for shunting off all the positive charge encounters some obstructing configuration. Moreover, one can in principle enumerate the set of all such obstructing configurations which could arise when applying the given algorithm. The set of obstructing configurations is unavoidable; any maximal planar graph must contain one or else it could be discharged by the algorithm.

Now we shall make these ideas more precise and prove that some charge does indeed exist. The *initial charge* associates to every vertex v in a triangulation the integer $\xi_0(v) = 6 - \deg(v)$.

Lemma 3-2 The sum of the initial charges is 12.

PROOF Let p be the number of vertices and q the number of edges in the triangulation; then

$$\sum_v (6 - \deg(v)) = 6p - \sum_v \deg(v)$$
$$= 6p - 2q$$

By Euler's formula, $q = 3p - 6$ for triangulations, as we saw in Sec. 3-2, and so $6p - 2q = 12$.

In view of this lemma, let us call a *charge* any function ξ which assigns a number $\xi(v)$ to each vertex so that $\sum_v \xi(v) = 12$. Charge is redistributed in a triangulation by moving it away from some vertices to other vertices so that the total charge is preserved. A *discharging algorithm* is a rule for redistributing the charge in any triangulation. Since the rule is to apply to any triangulation, it must be specified solely in terms of local information. For instance, the rule might tell us to send the initial charge of 1 at a vertex of degree 5 to its various major neighbors, splitting it equally among them. The algorithm is said to *discharge* a triangulation G if applying it to the initial charge ξ_0 produces a new distribution ξ_1 whose value $\xi_1(v)$ is nonpositive at each vertex v. Since any discharging algorithm preserves total charge, the total charge is still positive and so some vertex must have positive charge. Therefore, no discharging algorithm can discharge a triangulation. Some configuration of vertices must be present which prevents discharging from taking place. For instance, a vertex of degree 5 with no major neighbors will not be discharged by the preceding discharging rule.

As another example of discharging, we shall give a discharging algorithm due to Haken (1973a) and show how it discharges any triangulation G which fails to contain one of a short explicit list of configurations, called the *unavoidable set* produced by the discharging algorithm. In the next section, we shall discuss the production of unavoidable sets, all of whose elements are likely to reduce. That requires a much more complicated discharging algorithm, but for now our simple example illustrates the ideas of discharging and unavoidable sets. In particular, since the ultimate goal is to find an unavoidable set of reducible configurations, we do not hesitate to exclude the known reducible ones early.

Now for Haken's (1973a) algorithm. Consider an arbitrary triangulation G. We shall distribute the initial negative charge $6 - j$ of any vertex v_j of degree $j \geq 8$ to its five-vertex neighbors. The restriction $j \geq 8$ is a technical device to facilitate analysis of the obstructing configurations. The neighbors of v_j induce a j-circuit K_j in G. We claim that no chord of K_j is present; otherwise, G contains a separating three-circuit which we have already shown to be reducible. Hence, vertices in K_j are adjacent only when they are consecutive neighbors of v_j. We also assume that G has no vertices of degree < 5, since such vertices are reducible.

Each vertex in K_j has exactly two neighbors, and, in particular, a v_5 in K_j has zero, one, or two v_5-neighbors in K_j. If there are zero or two of these neighbors, assign to v_5 the weight factor $w = 2$ with respect to v_j. If v_5 has precisely one degree-5 neighbor, v_5 gets weight $w = 1$. Let p_1, p_2 denote the number of five-vertices in K_j with weights $w = 1, 2$, respectively.

If v_j has at least 1 five-neighbor, we distribute the negative charge $6 - j$ from v_j to each of its five-vertex neighbors, giving each v_5 adjacent to v_j a share corresponding to its weight. Thus, each v_5 in K_j receives an additional negative charge of $[w/(p_1 + 2p_2)](6 - j)$. If v_j has no five-neighbor, we do not redistribute the negative charge at v_j. This procedure will clearly discharge every vertex of degree ≥ 8. Call the new charge ξ_1.

By our construction $\xi_1(v_j) \leq 0$ for $j \geq 8$. We are about to show that the five-vertices are also discharged by the process unless certain configurations occur in the triangulation. But how will this algorithm succeed in discharging a vertex of degree 7? Obviously, it will not; and so, the first configuration which must be avoided by the triangulation is a vertex of degree 7.

To show that every v_5 is discharged, i.e., that $\xi_1(v_5) \leq 0$, we need to first estimate how much negative charge each five-vertex will be receiving from its neighbors of degree ≥ 8. Since our intention here is to demonstrate discharging, we shall make another simplifying assumption; namely, that there are no vertices of degree 6.

Suppose for a moment that v_5 has only one major neighbor. Then the first neighborhood of v_5 must be as shown in Fig. 3-13, in which "$\boxed{\cdot}$" denotes a vertex of unspecified degree ≥ 8. But this configuration contains the configuration B which we showed reducible in Sec. 3-3 (see Fig. 3-14). In fact, if v_5 has three consecutive five-neighbors, then its first neighborhood contains the configuration B.

Figure 3-13

B

Figure 3-14

Let us assume now that B is excluded from G (along with vertices of degree 6 or 7). Then we can prove a useful lemma.

Lemma 3-3 Every v_5 in G has at least two neighbors of degree ≥ 8. If there are exactly two such neighbors, then they are not adjacent to each other.

PROOF If v_5 has fewer than two neighbors of degree ≥ 8 or if v_5 has exactly two such neighbors which are consecutive, then by the exclusion of six- and seven-vertices, the remaining neighbors of v_5 include three consecutive five-vertices. But this is the configuration B which we have banished.

According to this lemma, there are three cases to be considered (see Fig. 3-15).

(1) v_5 has precisely two major neighbors and has weight $w = 2$ with respect to each of them.
(2) v_5 has precisely three major neighbors and has weight $w = 2$ with respect to exactly one of them.
(3) v_5 has at least four major neighbors.

If we could show that the (negative) charge transferred from v_j ($j \geq 8$) to each of its v_5-neighbors is at least $w/4$, where w is the weight of the neighbor with respect to v_j, then in each of the three cases (1), (2), and (3) above, the v_5 is sure to get enough negative charge from its major neighbors to be discharged. To insure that each v_5 does receive enough negative charge to have its initial positive charge nullified, we must exclude another configuration from G—a vertex denoted "□" of degree 8 with five consecutive v_5-neighbors (see Fig. 3-16). We also exclude three other configurations, originally considered by Birkhoff, con-

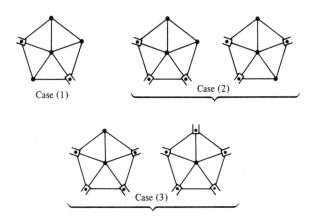

Case (1)

Case (2)

Case (3)

Figure 3-15

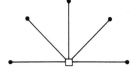

Figure 3-16

sisting of a vertex of degree j with $j - 1$ v_5-neighbors, where $j = 9$, 10, or 11. With these assumptions about G, we can now prove the critical lemma which guarantees complete discharging of G.

Lemma 3-4 If v_j is a vertex of G ($j \geq 8$) and if v_5 is a neighbor of v_j with weight factor w in K_j, then the negative charge t which is transferred from v_j to v_5 has absolute value $|t| \geq w/4$.

PROOF We divide the proof into three cases.

Case (1) $j > 11$. Recall that p_i denotes the number of v_5-neighbors of v_j getting weight factor i ($i = 1, 2$). Hence, $p_1 + p_2 \leq j$ and so, of course, $p_2 \leq j$. The charge t transferred from v_j to v_5 is $w[(6 - j)/(p_1 + 2p_2)]$ so

$$|t| = w \frac{j - 6}{p_1 + 2p_2} \geq w \frac{j - 6}{2j} \geq w \times \frac{1}{4}$$

The last inequality uses our assumption $j \geq 12$.

Case (2) $j = 9$, 10, or 11. Since G contains no v_j with $j - 1$ consecutive v_5-neighbors ($j = 9$, 10, or 11), we have $p_1 + p_2 \leq j - 2$. In K_j, the vertices of degree 5 (in G) induce a subgraph F_j consisting of isolated points and paths. The endpoints of the paths in F_j are just the v_5-neighbors of v_j which are assigned the weight factor $w = 1$. Thus, p_1 is even. If $p_1 = 0$, then F_j consists only of isolated points, in which case $p_2 \leq j/2 \leq j - 4$ for $j = 9$, 10, or 11. On the other hand, if $p_1 \geq 2$, then $p_2 = p_1 + p_2 - p_1 \leq p_1 + p_2 - 2 \leq j - 4$. Arguing as before, we find that the magnitude $|t|$ of the negative charge transferred from v_j to v_5 satisfies

$$|t| = w \frac{j - 6}{p_1 + 2p_2} \geq w \frac{j - 6}{2j - 6} \geq \frac{w}{4}$$

Case (3) $j = 8$. Since the configuration shown in Fig. 3-16 has been excluded, each component of the graph F_j has at most four vertices (or else v_8 would have five consecutive v_5-neighbors) and, hence, $p_1 + p_2 \leq 6$. If $p_1 + p_2 = 6$, the components of F_j must contain four and two vertices or three and three vertices, respectively. Either way, $p_2 = 2$ and, thus, $p_1 + 2p_2 = 8$. If $p_1 + p_2 = 5$, we could have (4, 1), (3, 2), (3, 1, 1), or (2, 2, 1) as the sequence of the number of vertices in each component of F_j. In each possible case, $p_2 \leq 3$ and so $p_1 + 2p_2 \leq 8$. If $p_1 + p_2 \leq 4$, then $p_2 \leq 4$ and, therefore,

$p_1 + 2p_2 \leq 8$. Now we compute the magnitude of the charge transferred:

$$|t| = w\frac{8-6}{p_1 + 2p_2} \geq w \times \frac{2}{8} = \frac{w}{4}$$

This completes the proof of the lemma since $|t| \geq w/4$ in all possible cases.

Let us summarize now. We have shown that a triangulation G, which contains no vertices of degree <5 and no separating three-circuits, can be discharged unless one of the following configurations is present: (*a*) a vertex of degree 6; (*b*) a vertex of degree 7; (*c*) a vertex of degree 5 with three consecutive v_5-neighbors (see Fig. 3-14); (*d*) a vertex of degree 8 with five consecutive v_5-neighbors (see Fig. 3-16); (*e*) a vertex of degree $j = 9, 10$, or 11 with $j - 1$ consecutive v_5-neighbors. Since no triangulation can be discharged, these configurations constitute the unavoidable set produced by Haken's algorithm. Moreover, all except the first two are known to be reducible. It follows from this that any minimal counter-example to the 4CC, had there been one, would have contained a vertex of degree 6 or 7. Otherwise, its four-colorability could have been deduced from the four-colorability of a graph with fewer vertices.

The *principle of discharging* (due to Heesch, 1969) takes a maximal planar graph T containing no member of such a finite set L of configurations and simply redistributes the charges, maintaining the restriction that the sum of the charges is 12 so that (*a*) all positive charges disappear, which contradiction shows that such a graph T cannot exist, or (*b*) no positive charge exceeds $s \leq 1$, from which it follows that T must have at least $12s^{-1}$ vertices. To distinguish between the two similar but not identical notions, we call (*a*) *discharging* and (*b*) *charge leaking*.

Heesch calls a set L of configurations *unavoidable* if any planar graph G must contain some member of L. As above, a set of configurations which prevents discharging is unavoidable. Some unavoidable sets L exist; for example, L can consist of single vertices of degrees 4 and 5. If the planar graph G is already known to avoid basic reducible configurations such as vertices of degree less than 5, then L consists of one configuration—a vertex of degree 5. Heesch hoped to prove the 4CC by finding an unavoidable set in which every configuration is reducible.

If L consists of configurations known not to occur in a minimal five-chromatic graph G, then using charge leaking for G produces lower bounds on the Birkhoff number B which is the number of vertices in G. Presumably, the longer the list of configurations in L, the higher we can raise B.

As an illustration of the theory, let us prove $B \geq 26$. Let L be the following set of configurations: $L = \{l_1, l_2, l_3\}$, where l_1 is a minor vertex with three consecutive five-vertex neighbors; l_2 is a major vertex with all neighbors but one of degree 5; and l_3 is a five-vertex with all minor neighbors.

Theorem 3-17 Let G be a maximal planar graph with $\delta(G) = 5$ which avoids L. Then G has at least 26 vertices.

Since the configurations l_1, l_2, and l_3 are all excluded from minimal graphs by Theorems 3-10 and 3-11. Corollary 3-1, and Theorem 3-12, respectively, we have the following corollary.

Corollary 3-2 $B \geq 26$.

PROOF OF THEOREM 3-17 (We follow an argument of Bernhart, 1974a.) We shall redistribute the initial charge so that no vertex has charge exceeding $s = \frac{12}{25} = \frac{48}{100}$ and some vertex has charge less than s, from which it follows immediately that there must be *more* than $12s^{-1} = 25$ vertices. Hence, G has at least 26 vertices.

Now for the distribution of charge. We take charge from the vertices of degree 5 and distribute it to their neighbors of degree > 5.

(1) If there are ≥ 2 major neighbors, give $\frac{26}{100}$ each to two of them.
(2) If there is one major neighbor, give it $\frac{28}{100}$ and give $\frac{12}{100}$ apiece to two neighbors of degree 6 (if possible).

Let us show first that no vertex v of degree 5 has charge $q(v)$ greater than $\frac{48}{100}$. If case (1) holds, then $q(v) = 1 - 2 \times \frac{26}{100} = \frac{48}{100}$. If case (2) is applicable, note that the four nonmajor neighbors of v cannot all be of degree 5; in fact, either two of them have degree 6 immediately or, in cyclic order, they must have degrees 5, 6, 5, 5 because G does not contain l_1. But then the neighbor of degree 6 has three consecutive neighbors of degree 5, which is again impossible since $l_1 \in L$. Hence, if v has only one major neighbor, it also has at least two neighbors of degree 6. Thus, $q(v) = 1 - \frac{28}{100} - 2\frac{12}{100} = \frac{48}{100}$. Because $l_3 \in L$, we have considered all vertices of degree 5.

Now suppose w has degree 6. By l_1, w has at most four neighbors of degree 5 so $q(w) \leq 0 + 4\frac{12}{100} = \frac{48}{100}$.

Finally, suppose degree $x = k \geq 7$. By l_2, x has at most $k - 2$ neighbors of degree 5. Hence,

$$q(x) \leq (6 - k) + (k - 2)\frac{28}{100} = \phi(k)$$

Now $\phi(7) = \frac{2}{5}$ and $\phi(k_1) < \phi(k_2)$ if $k_1 > k_2$. Hence $q(x) \leq \frac{2}{5} < \frac{48}{100}$. This completes the argument since, by l_3, G has some major vertex.

Until recently, the best-known bound on B was $B \geq 40$ due to Ore and Stemple (1970). Their argument was very lengthy and the published proof was merely a summary of the method. However, Mayer (1969) has obtained an elegant argument to show $B \geq 48$, and we shall give a very quick sketch of his argument.

Theorem 3-18 $B \geq 48$.

SKETCH OF PROOF Let n_i be the number of vertices of degree i and let $M = n_7 + n_8 + \cdots$ be the number of major vertices. Mayer shows

(a) $2n_6 + n_7 \geq 24 + n_9 + \sum_{i > 9} (i - 9)n_i$

(b) $M \geq 7$, with $M > 7$ if $M = n_7$

(c) $n_6 + n_7 + n_8 \geq 28$

Now (b) and Theorem 3-17 imply $n_5 \geq 20$. Adding (c) yields $B \geq 20 + 28 = 48$. The result (a) generalizes a theorem of Choinacki (1942):

(a') $3n_6 + n_7 \geq 24 + n_9 + \sum_{i > 9} (i - 9)n_i$

and is used in the derivation of (b) and (c).

Mayer's method is to carefully leak charges using a variety of somewhat complicated reducible configurations. One advantage is that his proof is self-contained except for reducible configurations. Before Haken and Appel found their proof, Mayer had used these techniques to push the lower bound on B up to 96.

3-5 COMPUTER-ASSISTED REDUCIBILITY

Appel and Haken appear to have completed the evolution of the proof of the four-color theorem. They first obtained reasonable criteria for the likely reducibility of configurations and then modified Heesch's original discharging algorithm so that the unavoidable sets produced contained only configurations which were likely to reduce. The next step, in which Appel and Haken were assisted by Koch, involved actually testing these configurations for reducibility. When certain configurations were found that could not be reduced with the available techniques, the discharging procedure was altered to produce better unavoidable sets which did not contain these configurations. The unavoidable set was also tailored to fit the abilities of computer-implemented reducibility algorithms which were, in turn, adapted from the algorithms of Heesch. Conversely, their reducibility algorithm took advantage of the flexibility of their discharging procedure (see Fig. 3-17).

The actual testing process for reducibility represented the final hurdle in the proof of the four-color theorem. Literally years of high-speed computer calculations were required to demonstrate that every member in an increasingly extensive list of configurations was reducible. This was a delicate matter. Had the list been much longer, had the specific reducibility checks required more time (some took hours), either because of inefficient programming or because of the innate complexity of the configuration, had the computer itself been too slow—had any of of these things happened—the total computer time might have measured decades instead of years and the proof probably would never have been found.

The construction of such an elaborate argument is a tribute to man's intuition, to his problem-solving ability and to his perseverence. Much more important than the initial stimulus, the computer-assisted analysis of cases employed by

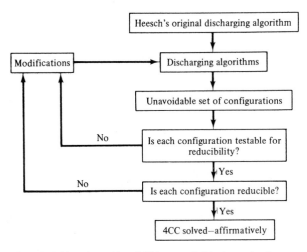

Figure 3-17 Flow chart of Appel–Haken proof of the four-color theorem.

Haken and Appel may lead to fundamental changes in our perception of mathematices and its role in solving complex problems. The symbiosis between mathematician and machine required for this proof points toward an exciting future.

As we have said before, in 1969 Heesch had already constructed the requisite machinery: discharging and unavoidable sets. He had also begun to investigate conditions under which configurations were likely to be reducible and believed that it would be possible to find a finite unavoidable set of reducible configurations, each of which could be found within the second neighborhood of a pair of vertices.

In 1971, Heesch contributed another key observation; he noted three "reduction obstacles" whose presence inhibits the reducibility of a configuration. A configuration R is called *geographically good* if it does not contain either of the first two of these obstacles:

(1) a four-legger vertex—i.e., a vertex of R connected with four or more vertices in the ring Q surrounding R;

(2) a three-legger articulation vertex—i.e., a vertex whose removal separates R and which is connected to three or more vertices in Q.

Avoiding these two obstacles is not quite enough to insure reducibility of a configuration. For example, the *triad*, which consists of three v_5-vertices forming a triangle, is geographically good but was shown to be nonreducible by Bernhart (1947). The remaining obstacle eliminates the triad:

(3) a hanging 5–5 pair—i.e., a pair of adjacent five-vertices connected by edges to only one other vertex of the configuration.

Configurations which avoid these three reduction obstacles are termed *likely to reduce* for the following heuristic reason: the presence of one of these subconfigurations appears to prevent the reducibility of the configuration (unless it contains a subconfiguration which is already reducible and does not contain the obstacle).

Appel and Haken saw that they could use these ideas to develop a systematic method for trying to prove the 4CC; let us call it *computer-assisted reducibility.* Begin with the Heesch discharging algorithm or some variant. Now try to manipulate the discharging procedure so that all the configurations in the unavoidable set which it produces are likely to be reducible in the above technical sense. When such an unavoidable set is found, test each of its elements using computer-implemented algorithms to show either directly that it is reducible or else that it contains a configuration which was previously proved reducible. If some configuration in the unavoidable set cannot be shown to be reducible, in spite of satisfying Heesch's three criteria for likely reducibility, go back to the discharging algorithm and change it again to replace the offending configuration by other likely-to-reduce configurations. But how do we know that computer-assisted reducibility will work? What if this process simply produces larger and larger unavoidable sets? Why do we expect to find an unavoidable set in which every configuration is reducible?

Before we can go further in our discussion, we must examine the nature of reducibility. So far, we have been content to define reducibility in a purely theoretical fashion. But to be implemented on the computer, an algorithm is needed which can definitely establish reducibility in a finite number of steps. Fortunately, there is an appropriate notion available; it is that of D-reducibility.

The kind of configuration which we shall be trying to reduce consists of a specifically embedded planar graph R, all of whose finite regions are triangular and whose vertices have each been prescribed a degree. Using the degree constraints on the boundary vertices of R and the elementary properties of reducible configurations already developed (such as the nonexistence of separating three-circuits), we can extend R to an enlarged configuration $\bar{R}$ which must appear whenever R does. Note that $\bar{R}-R$ is the circuit Q induced in any triangulation G containing R by the vertices of G adjacent to R but not in it. The number of vertices in Q is called the *ring size* of R and is often denoted by n.

For example, suppose R consists of a v_5 with all v_5-neighbors. Then Q is a circuit of length 5, and the resulting $\bar{R}$ is shown in Fig. 3-18.

If a configuration R is embedded in some triangulation G, the *complementary graph* $U = G-R$ has one nontriangular region which is bounded by the circuit $Q = \bar{R}-R$. If U has a four-coloring c, then a four-coloring c' is induced on Q. This four-coloring of Q may extend directly to a four-coloring of all of $\bar{R}$. In this case, R is certainly reducible.

However, we shall reserve the term D-reducibility for a more general concept. Suppose that for any four-coloring c of any complementary configuration U, the induced four-coloring c' of Q is either directly extendable to $\bar{R}$ or else c' is not extendable but can be modified by interchanging colors in the Kempe chains of c to produce another four-coloring c'' of Q which is extendable to $\bar{R}$. In this case, R is called D-reducible.

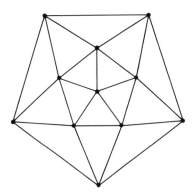

Figure 3-18

For example, if R consists of v_4, then Q is a four-circuit. If the four-coloring c' of Q uses less than all four of the colors, then c' extends directly. On the other hand, if the coloring c' displays all four colors on Q, then we have seen that two nonadjacent vertices of Q are joined by a Kempe chain (corresponding to one color pair—say rb) in c and the remaining two belong to distinct Kempe chains (of the opposite color pair—gy in this instance). Reversing the colors in either of these latter two Kempe chains changes the color of exactly one vertex of Q and thus converts c' into a four-coloring c'' which is extendable to v_4. This argument shows that v_4 is D-reducible.

Let us examine this Kempe-chain aspect of D-reducibility a little more carefully. If we begin with a four-coloring c' of Q, then each partition of the four colors into two complementary pairs induces a partition of the vertices of Q into the Kempe chains in c'. Given a complementary configuration U and a four-coloring c of U inducing c' on Q (such a coloring of U is called *feasible*), the color partition we have chosen for Q determines Kempe chains of c in U, each of which intersects Q in a union of Kempe chains of c' called a *Kempe residue*. Applying a Kempe-chain reversal to c in order to modify c' must reverse the colors in all of the Kempe chains of c' belonging to a single Kempe residue.

Now consider the problem of finding an algorithm for checking D-reducibility of a configuration R. The bounding circuit Q may have a fairly large number of different four-colorings (e.g., if Q has 13 vertices, there are 66,430 different four-colorings). But these numbers are finite and not excessively big for a modern high-speed computer. Moreover, the four-colorings of Q can be systematically generated. Given a particular four-coloring c' of Q, we can try to extend it to all of R. This might be done by first enumerating all possible four-colorings of $\overline{R}$ and listing the induced four-colorings of Q. An arbitrary four-coloring of Q extends to $\overline{R}$ if and only if it belongs to this list.

But what if the coloring c' of Q does not extend? In that case, for at least one color partition, we must consider the finite number of potential arrangements of Kempe chains of c' into Kempe residues, and for each such arrangement, we

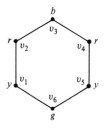

Figure 3-19

must find an allowable color reversal which converts c' into an extendable coloring. This shows that, regardless of how R is embedded in a triangulation G so that the complementary configuration U has a feasible four-coloring c (necessarily inducing c' on Q), there is a sequence of Kempe-chain reversals in c which converts c' into an extendable coloring.

The best way to understand this is to look at a simple example. Suppose that Q consists of a six-circuit which is four-colored as in Fig. 3-19 above.

If we partition the four colors into the pairs r, b and y, g then there are only two Kempe chains in c'. The first consists of $\{v_2, v_3, v_4\}$ and the second of $\{v_1, v_6, v_5\}$. Reversing the coloring of either or both of these Kempe chains does not produce an essentially different coloring of Q. For example, reversing the coloring of $\{v_2, v_3, v_4\}$ amounts to simply interchanging the roles of r and b in the coloring.

On the other hand, the partition rg/by yields six Kempe chains, one for each vertex. These may be joined to one another in various ways by Kempe chains in any feasible c.

If v_2, v_4, and v_6 all belong to the same rg component of U, then necessarily v_1, v_3, and v_5 are each in separate by-components. This situation can be described by a tree called by Tutte and Whitney a *chromodendron* (see Fig. 3-20), each vertex of which corresponds to a Kempe residue.

If $\{v_2, v_4\}$ and $\{v_1, v_5\}$ are Kempe residues, then so are $\{v_3\}$ and $\{v_6\}$ and the related chromodendron is as shown in Fig. 3-21.

Altering the coloring c' by reversing the colors on some sequence of Kempe residues in Q amounts to choosing a subset of the vertices of the tree. For

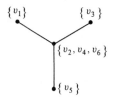

Figure 3-20

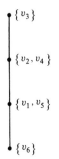

Figure 3-21

instance, choosing the vertex $\{v_3\}$ of the above chromodendron means changing b to y at vertex $\{v_3\}$ in the coloring c' of Fig. 3-19. The resulting coloring is quite different from c', having one color repeated three times (see Fig. 3-22).

Let us summarize our D-reducibility algorithm for a configuration R. We must check for each four-coloring c' of Q:

(1) Does c' extend to $\overline{R}$?

If not, then choose a color partition and check for all possible Kempe-residue structures relating the different Kempe chains of c'.

(2) Does there exist a sequence of Kempe-residue reversals carrying c' to some coloring c'' which does extend?

If at least one of these two questions has a "yes" answer for each c', then R is D-reducible. To carry out (2) systematically, we need to explore the theory of chromodendra more thoroughly.

Recall that the intersection of a Kempe chain of c with Q is called a Kempe residue of c in c', and each Kempe residue is a union of Kempe chains of c'.

Let $R(\pi)$ denote the set of all Kempe residues of c belonging to the partition $\pi = \alpha\beta/\gamma\delta$. Thus, an element r of $R(\pi)$ consists of a union of $\alpha\beta$ (or $\gamma\delta$) chains in Q which are "joined" by an $\alpha\beta$ (or $\gamma\delta$) chain $K(r)$ of U. (Note that by connectedness, $K(r)$ is uniquely determined by r.)

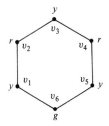

Figure 3-22

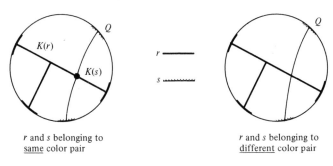

r and s belonging to
<u>same</u> color pair

r and s belonging to
<u>different</u> color pair

Figure 3-23

By the planarity of U, one member r of $R(\pi)$ cannot separate in Q two parts of another member s of $R(\pi)$, for if both r and s belong to the same color pair, say (α, β), then the hypothesis of separation would force $K(r) \cap K(s) \neq \varnothing$; so $K(r) = K(s)$ and hence $r = s$. On the other hand, if r and s belong to different color pairs, then $K(r) \cap K(s) = \varnothing$. But again the hypotheses of separation would imply $K(r) \cap K(s) \neq \varnothing$ (see Fig. 3-23).

In view of this observation, we see that if r, $s \in R(\pi)$, then they are separated from each other by two intervals I and J in Q, as indicated in Fig. 3-24.

Suppose I consists of a single edge. Then certainly r and s correspond to complementary color pairs, say $\alpha\beta$ and $\gamma\delta$, respectively, and in this case, Whitney and Tutte (1972) show that J also consists of a single edge, for if J contains any vertices, consider the vertex v separated from r by a single edge. By definition of a Kempe residue, v must belong to some $\gamma\delta$ chain s' and this $\gamma\delta$ chain must be separated from s in Q by an $\alpha\beta$ chain r'. It is now easy to deduce from the basic properties of Kempe chains that either r is joined to r' by an $\alpha\beta$ chain in c or s is joined to s' by a $\gamma\delta$ chain in c. Hence, either r or s would contain vertices of J violating our original assumption.

The foregoing remarks suggest a definition and a theorem. Define a graph $X(\pi)$ to have as vertices the Kempe residues in $R(\pi)$ and say that r, $s \in R(\pi)$ are adjacent when they are separated in Q by exactly two edges (which join them).

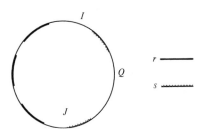

Figure 3-24

All of the vertices of $X(\pi)$ adjacent to a Kempe residue r are Kempe residues for the complementary color pair and any two are separated by r. Hence, every vertex of $X(\pi)$ with degree > 1 is a cut-point. Finally, note that, since no member of $R(\pi)$ separates another, $X(\pi)$ is connected. Thus we have proved the following theorem.

Theorem 3-19 $X(\pi)$ is a tree.

We call $X(\pi)$ a *chromodendron*. It is determined by the coloring c of U and the partition π; we write $X(c, \pi)$ when we wish to make this relationship explicit.

Figure 3-25 gives an example due to Tutte of a coloring c of U and a related chromodendron. For convenience, we have drawn U inside Q instead of outside.

Let us examine Fig. 3-25 to see how it illustrates the preceding remarks. The $\alpha\gamma$ chains AJ and H in Q are joined by a $\alpha\gamma$ chain in c and constitute a Kempe residue, represented by the vertex AJH of the chromodendron. Similarly, the vertex I represents the Kempe residue consisting solely of the $\beta\delta$ Kempe chain I. The edge of the chromodendron joining I and AJH in the chromodendron corresponds to the two edges IH and IJ of Q. The partition $\pi = \alpha\gamma/\beta\delta$ divides Q into 8 two-colored components (the $\alpha\gamma$ and $\beta\delta$ Kempe chains of c') which are amalgamated into five Kempe residues because of the arrangement of

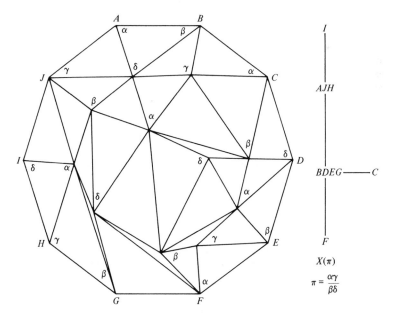

Figure 3-25

Kempe chains in c. Thus, the chromodendron $X(\pi)$ has five vertices and, since a tree has one less edge than vertices, there are four edges.

The key to systematizing our search for D-reducible configurations lies in the following two results. We shall prove that the number of vertices in any chromodendron $X(c, \pi)$ is determined by π alone. Then we observe that the number of essentially distinct Kempe-chain arrangements is just the number of different chromodendra and that this number is finite, depending only on c' and π.

For any partition π of the four colors into two complementary pairs, say $\pi = \alpha\beta/\gamma\delta$, the circuit Q is split into $\alpha\beta$ and $\gamma\delta$ Kempe chains. Let $q(\pi)$ denote the number of these two-colored components.

Lemma 3-5 If c' is any coloring of Q, c any feasible coloring of a complementary configuration, and π a partition of the four colors, then

    (a) $q(\pi)$ is even;

    (b) $X(c, \pi)$ has $\frac{1}{2}q(\pi)$ edges and $\frac{1}{2}q(\pi) + 1$ vertices.

PROOF Since the $q(\pi)$ different two-colored components appear in cyclic order around Q, $q(\pi)$ is even and the Kempe chains of c' must alternate, with $\frac{1}{2}q(\pi)$ belonging to each pair of the partition (see Fig. 3-20 and the discussion immediately following). Moreover, there must be $q(\pi)$ edges of Q which join Kempe chains belonging to complementary color pairs.

When the Kempe chains of c' are amalgamated into Kempe residues, i.e., given a particular Kempe-chain arrangement for some feasible c, each edge of the resulting chromodendron is an adjacency between two Kempe residues and corresponds to two of the $q(\pi)$ edges of Q (see Fig. 3-24). Thus, the chromodendron $X(c, \pi)$ generated by any Kempe-chain arrangement relative to the partition π has $\frac{1}{2}q(\pi)$ edges and $\frac{1}{2}q(\pi) + 1$ vertices.

For a partition π and coloring c', calling two Kempe-chain arrangements essentially different means that they induce different amalgamations of the Kempe chains in c' into Kempe residues. But each such amalgamation produces a chromodendron. In other words, we may have $X(c_1, \pi) = X(c_2, \pi)$ only when c_1 and c_2 yield essentially the same Kempe-chain arrangements. Since each vertex of a chromodendron corresponds to a subset of the $\frac{1}{2}q(\pi)$ components of Q colored with one of the complementary color pairs and since the chromodendron has $\frac{1}{2}q(\pi) + 1$ vertices, there is plainly an upper bound on the number of chromodendra $X(\pi)$ in which $q(\pi)$ has some fixed value. Since $q(\pi)$ takes on at most a finite number of values, we have certainly established the next lemma.

Lemma 3-6 There are a finite number of chromodendra associated with any coloring c' of Q, and this is exactly the number of essentially different Kempe-chain arrangements in any feasible coloring c of a complementary configuration U.

We have not specifically described computational detail of the Appel–Haken proof (1976c). There is a good reason for this omission. To present all of the details of their discharging algorithm, we would have to reproduce the 26 pages of tables and 15 figures which they used to describe it. Essentially, the idea is simple enough. Two kinds of discharging are distinguished: (1) short-range (which are further subdivided into S and L discharging) and (2) transversal (or T) discharging. Charge is transferred from a v_5-vertex to a major vertex; in the first case, the charge flows directly along an edge, while the second-case charge flows transversally across 6–6 edges (which join v_6-vertices). For instance, Fig. 3-26, called T1 #1 by Appel and Haken, indicates that a charge of $\frac{1}{6}$ is to be transferred from v to w. The notation means that the degrees are as follows: $\deg(v) = 5$, $\deg(w) \geq 7$, $\deg(a) = 6 = deg(b)$, $\deg(c) \geq 6$, $\deg(d) \geq 6$.

Similarly, we have not duplicated the 63 full-page figures, each containing up to 35 separate configurations, which appear in the table of configurations in the unavoidable set. For the details, see Appel, Haken, and Koch (1976d).

We have given the logical structure of this computer-assisted proof with all the reducibility and discharging ideas. Because the implementation of their ideas involved such an elaborate program, not all mathematicians are convinced of its correctness. Indeed, small errors have been found, but they have always been correctable. In the next section we shall give the elegant but ad hoc arguments which convinced Appel and Haken that their method must work.

3-6 THE PROBABILITY ARGUMENT

The crux of the Appel–Haken proof of the four-color theorem is a subtle and elegant probabilistic argument which establishes an a priori certainty that there must exist some discharging procedure producing an unavoidable set all of whose configurations are reducible. That is, they showed that the computer-assisted reducibility proof was overwhelmingly likely to succeed, and they appear to have been right.

In a way, this sort of reasoning should be very familiar. As a preliminary to approaching any problem, one makes a brief (and somewhat automatic) estimate of the likelihood of various methods to yield a solution. When dealing with such a huge and elaborate proof technique as computer-assisted reducibility,

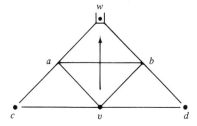

Figure 3-26

it becomes particularly important to verify a hunch in advance as carefully as possible.

The probability arguments to be given now are admittedly crude and make several unjustified assumptions. In retrospect, however, we know that they are valid in that they lead to a mathematically correct proof. Perhaps the best analogy is with the intuition which is required to obtain a new experimental result. Once the result has been found, one no longer needs to defend the likelihood that the experiment will succeed in demonstrating it. Since the "experiment" which Appel and Haken wished to perform would take years of valuable computer time, not to mention their own time and effort, even a crude argument was better than none.

There are really two stages to this process. The first lifts us off the ground and points us in the right direction, while the second stage takes us there. To begin with, we examine the qualitative behavior of the probability that a configuration is reducible. This analysis implies that for a configuration R of fixed ring size n, as the number m of vertices in R exceeds some critical value, R becomes very likely to reduce. The differential between the growth rate of area over circumference now suggests that for n sufficiently large, any R is likely to reduce.

This is the end of the first stage and it is vital information, for it shows that our discharging procedure will not require an arbitrarily large amount of modification. Once the algorithm is complex enough, the configurations which form the unavoidable set needed to describe all failure cases have ring sizes exceeding the critical value of n and are likely to reduce. This defeats the first major objection to a computer-assisted reducibility proof—that perhaps every unavoidable set will contain nonreducible configurations.

There is, however, a second major or potential obstacle. What if the critical value of n were, say, 100? How would we be able to check the reducibility of such configurations within a reasonable span of time? The second stage of the Haken–Appel plausibility argument settles this problem by estimating the critical value of n using a simple average-degree technique. In fact, they show, as we shall see, that the critical value of r is certainly ≤ 17 and probably ≤ 14. Since configurations of ring size 17 can be checked for reducibility via computer in the time frame which they had allowed themselves, Appel and Haken were now sure that their method would work.

Rather than using 17, they actually insisted that their configurations have ring size not exceeding 14. If the discharging procedure produced a configuration of ring size 15, the procedure was modified to yield only configurations of smaller ring size. Since the final unavoidable set of reducible configurations contained none with ring size greater than 14, the assumption was indeed justified. Incidentally, using 14 instead of 17 had the fringe benefit of decreasing the time required to run the final experiment from four or five years to six months.

It may be of interest to note that the time required for checking D-reducibility of a configuration is roughly proportional to the number of possible four-colorings of the bounding ring. This number of four-colorings is very large and grows by approximately a factor of 3 for each successive increment in the number of vertices in the bounding ring. Thus, the jump from 14 to 17 would have required substantially more time, and a ring size of 20, say, would have been totally impossible even with sophisticated present computing machinery such as the IBM 370-160A used extensively in the solution. Here are some of the numbers involved: rings of size 10 have 2,461 colorings, size 11 have 7,381, size 12 have 22,144, size 13 have 66,430, size 14 have 199,291, and size 15 have 597,871.

We shall now estimate the probability that a given configuration R is D-reducible. Let y denote the probability that R is D-reducible and x the probability that an arbitrary coloring c' of Q is directly extendable to a coloring of $\overline{R}$. As x increases, we shall prove that y increases quite sharply, becoming very near 1 by the time $x \geq 0.30$. The qualitative behavior of x, and hence of y, is then shown to be dependent on the geometry of the configuration.

Call a coloring c' of Q *initially good* if it can be directly extended to $\overline{R}$. If a coloring c' is not initially good, it is *ultimately good* provided that for any possible Kempe-chain arrangement in any feasible coloring c of a complementary configuration U, there is a sequence of Kempe-chain reversals which transforms c' into an initially good coloring.

Our first job is to translate the definition of ultimately good, given above, into a statement about chromodendra. As we have already seen, a Kempe-chain arrangement can be described by a chromodendron which is a graph with each vertex representing a Kempe residue and each edge representing adjacency of Kempe residues. To modify c' by reversing certain Kempe chains of c amounts to modifying c' by reversing the corresponding Kempe residues in c'. This can be done as follows. First, choose a partition π of the four colors, say $\pi = \alpha\gamma/\beta\delta$. Second, choose one of the chromodendra $X(c, \pi)$, where c is a feasible coloring. (As we have seen, there are a finite number of such chromodendra and this number depends only on c' and π.) Third, and finally, choose a subset of the vertices of $X(c, \pi)$ and reverse the colorings of the corresponding Kempe residues in c'.

This color reversal can be performed in $2^{\frac{1}{2}q(\pi)+1}$ ways since $X(c, \pi)$ has $\frac{1}{2}q(\pi) + 1$ vertices. However, only $2^{\frac{1}{2}q(\pi)-1}$ of these reversals yield essentially distinct colorings of Q. Two colorings of Q are *not* essentially distinct if they induce the same partition of Q's vertices into four color classes. Since there are two possible color pairs to reverse or not reverse, each coloring of Q belongs to a set of 2^2 not-essentially distinct colorings, which accounts for the preceding decrease in the number of essentially distinct color reversals by a factor of 4. Colorings of Q which are not essentially distinct have exactly the same extendability properties—one extends to $\overline{R}$ if and only if any other extends.

For instance, selecting the empty subset of vertices of $X(c, \pi)$ amounts to leaving c' alone. On the other hand, we could select only the vertices of $X(c, \pi)$ corresponding to $\alpha\gamma$ chains of c'. Reversing these chains simply has the effect of reversing the colors α and γ in c', which produces a coloring of Q not essentially distinct from c'. Plainly, reversing the $\beta\gamma$ chains or the $\alpha\gamma$ and $\beta\delta$ chains also produces colorings of Q not essentially distinct from c'.

If we begin with a coloring c' that is not initially good, and if we have some partition π with $q(\pi) = q$, then to prove c' ultimately good, we have $2^{\frac{1}{2}q-1} - 1$ different possible reversals (one less than $2^{\frac{1}{2}q-1}$ since c' is assumed not initially good and so the "null" reversal is excluded) for each of the chromodendra $X(c, \pi)$. A successful reversal—i.e., one converting c' into an initially good coloring —must be found for each chromodendron.

Now we can begin to estimate probabilities. Choose a coloring c' at random from the set of all four-colorings of Q. There is a probability x that c' is initially good and $1 - x$ that c' is not initially good. In the latter case, we want to determine the probability that, for each possible Kempe-chain arrangement, there is some set of Kempe-chain reversal that can convert c' into an initially good coloring.

Let us be specific. Suppose Q is a 13-ring. We shall estimate the probability that a four-coloring c' of Q, which is not initially good, can always be modified to form an initially good coloring, regardless of the arrangement of its Kempe chains into Kempe residues. It is clear that for any partition π of the four colors, $2 \leq q(\pi) \leq 12$. In a moment we shall prove that there exist distinct partitions π_1 and π_2 with $q(\pi_1) = 10$ or 12 and $q(\pi_2) \geq 8$. This justifies the consideration of three special cases: the partition π has (1) $q(\pi) = 8$; (2) $q(\pi) = 10$; (3) $q(\pi) = 12$. In case (1), the chromodendra to be examined all have four edges and five vertices. There are thus $2^3 - 1 = 7$ possible reversals for each of these chromodendra and the probability that all seven of these modifications fail to be initially good is $(1 - x)^7$, since the probability is $1 - x$ that an arbitrary coloring is not initially good and since we are assuming, by default, that the respective probabilities are independent. Hence, the probability that at least one such modification of c' is successful is $1 - (1 - x)^7$. But this must hold for all of the possible chromodendra with four edges. Later, we will prove that there are exactly 14 such chromodendra and, therefore (again assuming independence), the probability, denoted y_4 by Appel and Haken, that in case (1) c' is ultimately good is

$$y_4 = [1 - (1 - x)^7]^{14}$$

If y_5 and y_6 denote, respectively, the probabilities that c' is ultimately good in case (2) or case (3), then since the number of chromodendra in these cases is 42 and 132, respectively, we have

$$y_5 = [1 - (1 - x)^{15}]^{42}$$
$$y_6 = [1 - (1 - x)^{31}]^{132}$$

These functions are graphed in Fig. 3-27.

At this point, we can return to the problem of estimating y, the probability that a configuration R of ring size 13 is D-reducible. An arbitrarily selected coloring c' of Q has probability x of being initially good and a probability of being ultimately good of at least $(1 - x)y_6$ if $q(\pi_1) = 12$ and at least $(1 - x)y_5$ if $q(\pi_1) = 10$. Thus, the probability y of being D-reducible is at least $x + (1 - x)y_5$ and possibly at least $x + (1 - x)y_6$. In fact, a careful reading of the

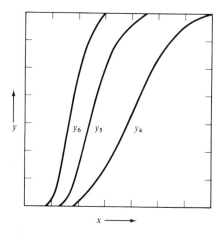

Figure 3-27

proof of Lemma 3-7 shows that the situation is even slightly better than we have described it. For example, whenever $q(\pi_1) = 10$ and $q(\pi_2) = 8$, the third partition π has $q(\pi) = 8$.

But we really have enough information as it is. The graphs given in Fig. 3-27 suggest that Kempe-chain arguments have very small chance when $x \leq 0.10$; very good chance when $x = 0.20$; and excellent chance for $x \geq 0.30$. Thus, as the probability x that a coloring c' of Q is initially good increases, the probability y that R is D-reducible rapidly approaches unity.

Of course, this is only a qualitative statement about y since estimating x seems to be out of the question. However, if we hold the ring size n constant and let the number m of vertices in R increase, it appears very reasonable that x should increase. This is because the number of colorings of R will increase and so the number of induced colorings of Q is also expected to increase. Hence, an arbitrary coloring of Q could be expected to become more likely to extend. We might, therefore, look for a specific critical value m' for which $m > m'$ means "very likely to be reducible" while $m < m'$ means "unlikely to be reducible."

We shall return to this question shortly, but first the lemmas promised earlier must be given. After considering the possible values of $q(\pi)$ for different partitions π, we will enumerate all of the chromodendra corresponding to partitions with $q(\pi) = 8$.

Lemma 3-7 If c' is any four-coloring of the 13-circuit Q, then there is some partition π_1 of the four colors with $q(\pi_1) = 10$ or 12 and there is another partition $\pi_2 \neq \pi_1$ with $q(\pi_2) \geq 8$.

PROOF Let $\pi = \alpha\beta/\gamma\delta$ denote the partition of the four colors in c' for which $q(\pi)$ is smallest. Since Q is not two-colorable, $q(\pi) \geq 2$ and certainly $q(\pi) \leq 12$.

In both of the remaining partitions π_1 and π_2 the colors α and β belong to distinct pairs, and similarly for γ and δ. This means that the edges of Q in an $\alpha\beta$ or $\gamma\delta$ chain must join distinct Kempe chains of c' for π_1 and π_2.

Suppose $q(\pi) = 2$. Then at most two edges of c' are not in $\alpha\beta$ or $\gamma\delta$ chains, and so $q(\pi_1) \geq 13 - 2 = 11$ and also $q(\pi_2) \geq 11$. Since q is even for any partition $q(\pi_1) = 12 = q(\pi_2)$, we are done when $q(\pi) = 2$.

Suppose $q(\pi) = 4$. Now at most four edges of c' will fail to separate Kempe chains in c' corresponding to π_1 or π_2. Hence, $q(\pi_i) \geq 13 - 4 = 9$ so $q(\pi_i) \geq 10$ for $i = 1, 2$. (In fact, the argument below actually shows $q(\pi_1) = 12$.)

The situation when $q(\pi) = 6$ is only slightly more complex. Arguing as before, $q(\pi_i) \geq 13 - 6 = 7$ so $q(\pi_i) \geq 8$ for $i = 1, 2$. Being slightly more careful, we notice that if one of the six questionable edges of c' fails to separate two π_1 chains, it certainly will separate two π_2 chains. Thus, in the worst case, at most three of these six edges fail to separate π_1 chains and so $q(\pi_1) \geq 13 - 3 = 10$. (Furthermore, if $q(\pi_1) = 10$, then $q(\pi_2) = 10$ also.)

Finally, if $q(\pi) = 8$, we can reason as in the preceding case. At most, four of the eight edges fail to separate π_1 chains so $q(\pi_1) \geq 13 - 4 = 9$ and hence $q(\pi_1) \geq 10$. Since $q(\pi) = 8$, we can take $\pi_2 = \pi$.

Now let us consider the number of different Kempe-chain arrangements.

Lemma 3-8 There are exactly 14 different chromodendra associated with a coloring c' of Q and a partition π for which $q(\pi) = 8$.

PROOF The three trees with four edges are

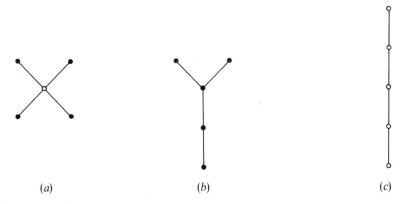

(a)   (b)   (c)

Let the Kempe chains of c be given in clockwise order by $K_1, \ldots, K_8$ and let $\pi = \alpha\beta/\gamma\delta$. Without loss of generality assume that K_1, K_3, K_5, K_7 are $\alpha\beta$ chains and that K_2, K_4, K_6, K_8 are $\gamma\delta$ chains. If all four $\alpha\beta$ chains are grouped together, in one Kempe residue, then the four $\gamma\delta$ chains must each represent a separate vertex of the chromodendron and we get and, reversing $\alpha\beta$ and $\gamma\delta$, we get

corresponding to the Kempe-chain arrangements

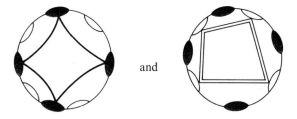

respectively. Since the Kempe-chain arrangements involving $\gamma\delta$ chains are completely determined by the arrangements involving $\alpha\beta$ chains, we could have described these latter two arrangements more succinctly by writing

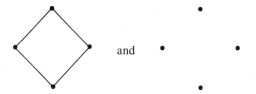

respectively. Using this notation, all of the Kempe-chain arrangements are given in Table 3-1. The total number is 14, as required.

Table 3-1

Total number	Kempe-chain arrangement				Chromodendron type
4					
4					
2					
2					
1					
1					

Similar arguments show that the number of chromodendra is 42 when $q(\pi) = 10$ and 132 when $q(\pi) = 12$.

Now we can return to the question of determining the critical m value m' for a given value of n so that any configuration R with $m > m'$ vertices and ring size n will be very likely to be D-reducible. To apply our probabilistic arguments, we must insist that R contain none of the known reduction obstacles. In particular, then, R must be geographically good without hanging 5–5 pairs.

Examination of these configurations, for ring sizes from 6 to 11, suggested that the critical m value is

$$m' = n - 5$$

It will be convenient to put this heuristic assumption in a slightly different form. Following Haken and Appel, let us define, for a configuration R with ring size n and m vertices,

$$\phi(R) = n - m - 3$$

Our previous hypothesis can now be stated in terms of ϕ: if R contains none of the three reduction obstacles given above and $\phi(R) \leq 1$, then R is likely to be D-reducible.

We would like to apply our assumption to an arbitrary R with $\phi \leq 1$. But then the problem is to know whether or not R contains a reduction obstacle. The first step in this direction was taken by Appel and Haken who showed that an arbitrary R satisfying $\phi(R) \leq 0$ must contain a geographically good subconfiguration R^* also satisfying $\phi(R^*) \leq 0$. However, R^* might still contain a hanging 5–5 pair.

In their solution of the four-color theorem, Appel and Haken took care of this problem with the following:

m-**Lemma** Let R be an arbitrary configuration of ring size n containing m vertices (inside the ring) and suppose R is embedded in a suitable graph G. If $\phi(R) < 3 - n/2$ (equivalently $m > \frac{3}{2}n - 6$), then R contains a geographically good subconfiguration without hanging pairs R^* with ring size n^* such that $\phi(R^*) < 3 - n^*/2$. Moreover, if A is an articulation vertex of R^* and if W_1 and W_2 are the two "wings" of A (produced by deleting A from R^*), then $m(W_i) = \frac{3}{2}n(W_i) - 6$ for $i = 1, 2$.

PROOF The above constraint forces $n \leq 5$ for $m = 2$ and 3 and $n \leq 4$ for $m = 1$. The latter case corresponds to a vertex of degree ≤ 4, which is impossible since G is suitable. The cases $m = 2$ and 3 both involve a nontrivial configuration R bounded by a ring Q of length $n \leq 5$. If $n \leq 4$, then the configuration Q is a separating circuit of length ≤ 4 and so is reducible. If $n = 5$, Q is also reducible—this time by Birkhoff's result (Theorem 3-9). When $m = 4$, $n \leq 6$ and there is exactly one possible configuration R, namely, the Birkhoff configuration B given by a five-vertex with three consecutuve five-neighbors. Of course, B is also reducible.

We shall prove the m-lemma by induction on m. Our previous arguments show that the lemma is satisfied vacuously for $m \leq 4$ since known reducible configurations are excluded from suitable graphs. This takes care of the basic step of the induction. Assume now that the m-lemma holds for all configurations of up to $m - 1$ vertices. We must show that it holds for R as well.

If R satisfies all of the requirements for R^*, we are done. If not, then R must contain one of the three reduction obstacles given at the beginning of Sec. 3-5. Suppose R contains a "four-legger" vertex v. Then deleting v from

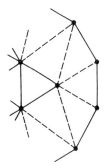 

Figure 3-28

R produces a new configuration R' with $m(R') = m - 1$ vertices. Moreover, the boundary ring Q' of R' replaces the chain of at least four vertices in Q to which v is adjacent by a chain of three vertices beginning at one end of the old chain, passing through v and then back out to the other end (see Fig. 3-28). Hence, $n(R') \leq n - 1$ so R' satisfies the inductive hypothesis and must contain an appropriate R^*. Very similar reasoning works also for the case of a "hanging 5–5 pair."

In the case that R contains an articulation vertex B, Haken and Appel show that at least one of the "wings" $W_1, \ldots, W_k$ ($k \geq 2$) at B satisfies the inductive hypothesis and so contains the required R^* (for $k > 2$; that is, for B a "three-legger" articulation vertex) while simultaneously showing that $m(W_i) = \frac{3}{2}n(W_i) - 6$ ($i = 1, 2$) when $k = 2$. This completes the proof.

Finally, to get an idea of how large n needs to be to insure likely reducibility, we introduce the notion of *size class* for neighborhoods in triangulations. Essentially, a neighborhood N_s of size class s is just the first neighborhood of a neighborhood N_{s-6} of size class $s - 6$. The following Table 3-2 explicitly describes N_s for $s < 8$ and gives the recursive relationship for $s \geq 8$. Each of the size classes in Table 3-2 has an "average n" and an "average m" (the average taken over all configurations of the size class in all triangulations).

For $1 \leq s \leq 5$, m is determined exactly in N_s ($m = s$ in these cases). However, n, the number of vertices adjacent to the neighborhood but not in it, is not exactly determined (nor is m for $s \geq 6$). For example, for size class 1, n is just the average degree of a vertex in any triangulation, averaged over all triangulations as well. Since the average degree of a vertex in a triangulation with p vertices is $6 - 12/p$ (as we proved earlier) the average value for n is 6 for size class 1. A similar argument shows that, for size class 2, the average value of n is 8. Since the average vertex degree over all triangulations is 6, we can assume that the edge joins two v_6-vertices, in which case one readily sees that $n = 8$.

This reasoning can be applied to the general case. Since the "average vertex

Table 3-2 (After Appel and Haken, 1976c.)

Size class	Description
1	Single vertex
2	Edge (pair of adjacent vertices)
3	Triangle (vertex with two consecutive neighbors)
4	Double triangle (vertex with three consecutive neighbors)
5	Triple triangle (vertex with four consecutive neighbors)
6	First neighborhood of a vertex (vertex with all neighbors)
7	Neighborhood N_6 of class 6 plus one triangle (with base in N_6)
8	First neighborhood of an edge
9	First neighborhood of a triangle
...	...
s	First neighborhood of a neighborhood N_{s-6} of size class $s - 6$ (for $s \geq 8$)
...	...

degree" is certainly 6 we may get an approximate idea of the average values for n and m by considering configurations (of the respective size classes) which consist of six vertices only. The corresponding ϕ values are plotted in Fig. 3-29 versus the n values and are marked X; the size class numbers are written above the marks. From size class 15 on ($n \geq 21$) the ϕ values lie below the line $\phi = 3 - n/2$, and thus the average configuration of this size class will almost certainly be D-reducible (see the *m-lemma*).

For example, for size class 6, we obtain average values for n and m, and hence, for ϕ, by considering the configuration consisting of a vertex of degree 6, all of whose neighbors have degree 6. This configuration has seven vertices (so the average values of m is 7) and ring size 12 (so the average n is 12). Since $12 - 7 - 3 = 2$, the average ϕ is 2 (see Fig. 3-30).

Every triangulation contains some configurations of each size class, the ϕ values of which are below the average since every triangulation contains vertices of degree 5 (i.e., with degrees substantially below the average of 6). For an estimate of these "unavoidable ϕ values" we have considered configurations of the different size classes such that one vertex, as close to the center as possible, is a five-vertex while all other vertices are of degree 6. The results are marked in Fig. 3-29. From $n = 17$ on, the values are in the region of extremely high likelihood of D-reducibility, since the unavoidable ϕ-vertices then lie below the line $\phi = 3 - n/2$.

In the case of N_6, for instance, we assume that the central vertex has degree 5. The corresponding configuration has $m = 6$ and ring size $n = 10$ so the respective averages are 6 and 10, and the average value of ϕ is 1.

It is easy to see, either by extrapolating the data given in Fig. 3-29 or by directly analyzing the average degree arguments above, that once below the line $\phi = 3 - n/2$, the unavoidable ϕ values will stay below. This is the justification for believing that the critical value of n is certainly ≤ 17. This means that the computer-assisted proof will not only probably work but will almost surely not require configurations with ring sizes exceeding 17. In fact, as we mentioned

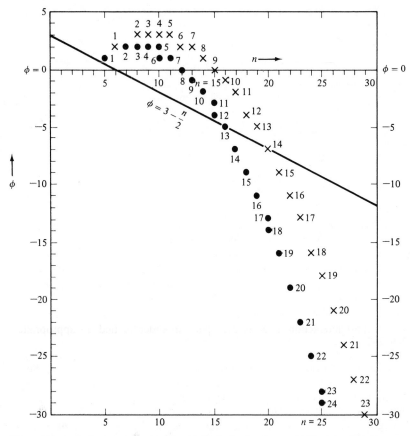

Figure 3-29 Approximate values of average ϕ (marked ×) and unavoidable ϕ (marked ●) versus n; size class as parameter. (*After Appel and Haken, 1976c.*)

before, Appel and Haken were actually able to keep $n \leq 14$. According to Fig. 3-29, the average ϕ values for $n = 14$ is $\phi = 1$, which is consistent with our earlier hypotheses on ϕ. Moreover, the unavoidable ϕ value for $n = 14$ is $\phi = -2$, which is very close to the critical line $\phi = 3 - n/2$.

3-7 CONCLUSION

There are several fundamental innovations in the Appel–Haken proof, but primary among these is the use of a computer to prove a mathematical theorem. Actually, the computer was used in several ways. To develop a discharging procedure which would yield an unavoidable set consisting only of likely-to-reduce

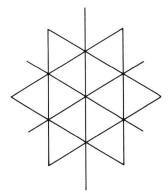

Figure 3-30

configurations, Appel and Haken began with an initial discharging algorithm and successively modified it. This involved considering the possible failure cases of each algorithm, which could occur because of the "overcharging" of some vertex; i.e., transferring to a vertex with negative charge too much positive charge. The enumeration of cases was done by computer, changes were made, and then the new algorithm was again examined. In other words, the computer was used via man–machine interaction as a "scratch pad" in order to find an appropriate discharging procedure.

The next step in the proof, checking configurations for reducibility, also employed the computer. We might call this sort of role that of an "idiot savant." While automated computing was a convenience which helped to speed up and systematize the search for the right discharging algorithm, it is absolutely necessary to test the D-reducibility of a configuration of ring size 14. Human computation is simply too slow.

The third point we wish to raise about computer involvement is that it was essential to the proof. In fact, any proof of the four-color theorem via Kempe chains and reducibility must be extremely complex, requiring computer assistance. This can be seen from the example given by Moore (Fig. 3-31) of a map in which there are no reducible configurations of ring size less than 11. It also appears quite unlikely that any of the equivalent formulations of the four-color problem, treated in the second part of this book, will yield a different solution (although the application of computer-assisted proof techniques may well be of benefit in these other areas).

We cannot totally rule out the possibility of a completely new proof utilizing geometric theorems more powerful than Euler's formula. However, Mark Thiel has shown that a considerable family of potentially distinct topological invariants all turn out to be polynomials in the Euler invariant and, thus, do not seem likely to yield new information or a new proof.

The chief difficulty, which had prevented use of the computer to solve the

North polar region—a nonagon

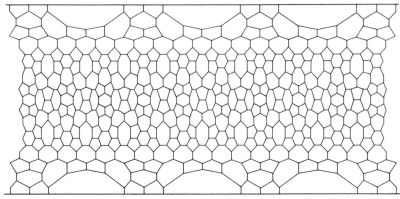

South polar region—a nonagon

Figure 3-31

4CC, was knowing exactly what questions to ask. As we mentioned in the previous section, Appel and Haken solved this problem with one of the most novel features of their proof. They devised a probabilistic argument which told them, fairly precisely, which configurations they could expect to reduce. Such a probabilistic approach not only showed that computer-assisted reducibility would very likely succeed but even guided its application.

This probability argument is the very sophisticated descendent of an old-fashioned hunch. It also demonstrates the central position of the mathematician in the Appel–Haken proof. Computer-assisted reducibility is a tool, allowing the computer to carry out the literally astronomical number of calculations required while it is aimed in the right direction by mathematical logic and probability theory. In fact, the Appel–Haken methodology suggests a new paradigm for mathematics.

This paradigm includes the traditional elements of intuition and standard logic, as well as heuristic and probabilistic techniques combined with the high-order computational abilities of a modern computer. At the beginning, any problem requires an intuition. This might be generated, for example, by the computer-assisted display of a large number of special cases—possibly in a graphical or interactive context. Next, the intuition is used to design a logical approach to the problem, including a means for modifying the approach in a systematic manner if it does not immediately yield success. This step may also involve interaction with the computer. Theorems and heuristic arguments are then advanced to estimate the probability that these logical procedures will reach their goal within a reasonable span of time. Finally, assuming that the probability estimate is strongly favorable, the computer experiment is performed and (hopefully) success is attained.

Let us extend these considerations one step further. Barring some basic error in either the discharging procedure or the reducibility checks (and such an error is highly unlikely to exist both for the reasons we have already indicated and also because of the extensive cross-checking and verifications now taking place), we shall be able to rely on the Appel–Haken proof and to call the four-color theorem a "theorem" in the classical sense. There are, however, problems of current mathematical importance for which no exact answer can be effectively computed. A large category of such examples is given by the notion of *NP*-completeness of problems (see Cook, 1971, and Karp, 1972). Roughly speaking, a problem is *NP*-complete if it is equivalent to any one of a growing list of important combinatorial problems—including the traveling salesman problem in which it is required to visit each of a set of cities exactly once with minimum total distance. A polynomial-time algorithm for one implies a polynomial-time algorithm for any other. The current opinion is that no such algorithms exist.

What this means, from a practical point of view, is that it may be effectively impossible to find, in the case of the traveling salesman problem, a shortest route through all *n* cities once *n* is sufficiently large. Aside from heuristics, which yield answers close to optimum in small cases but have no provable behavior for large *n*, the usual approach has been to prove approximation theorems. For example, one might be content to demonstrate the existence of a polynomial-time algorithm which yields a route not exceeding optimality by more than 50 percent. (See Christofides.)

A probabilistic approach, on the other hand, would consist of a polynomial-time algorithm and a theorem proving that the route provided by the algorithm has a probability of at least 95 percent, say, of being optimal. No such result is known (yet) for the traveling salesman problem. However, there is another problem, equally bad or even worse, for which such a theorem has been proved.

An integer p, greater than 1, is a prime if 1 and p are the only divisors of p. While it has been known since Euclid that arbitrarily large primes exist, no one has ever found a systematic procedure for constructing them. In fact, it is extremely difficult to determine whether a given large integer p is prime or not. For p sufficiently large, for example, $2^{400} - 593$, there may be no effective way to verify whether or not p is prime. It would simply take too long.

In such instances, when we cannot directly test whether success has occurred, we may have to be content with a high probability of success. Rabin (1976) has obtained a test for prospective primes p such that if n randomly selected numbers $k_1, \ldots, k_n$ all pass the test, then the probability that p is *not* prime is only $(\frac{1}{2})^n$. This sequence of tests amounts to a computer experiment in which we permit the computer a small but nonzero possibility of erring. Just as in an ordinary scientific experiment, one expects the repeatability of the experiment to effectively eliminate (or at least to minimize) the probability of experimental error—except, of course, for those errors which might be introduced as artifacts of the experimental design. On a formal level, one may regard all mathematical proofs as thought experiments which contain a nonzero possibility of error. Well-known cases in the literature illustrate how such an error may be missed for years (sometimes because of the small number of times the experiment is repeated). Presumably, some are never found. Error in the experimental design itself may occur if certain sets of axioms turn out to be inconsistent. The basic axioms of arithmetic, for example, are *not* known to be consistent.

To use the computer as an essential tool in their proofs, mathematicians will be forced to give up hope of verifying proofs by hand, just as scientific observations made with a microscope or telescope do not admit direct tactile confirmation. By the same token, however, computer-assisted mathematical proof can reach a much larger range of phenomena. There is a price for this sort of knowledge. It cannot be absolute. But the loss of innocence has always entailed a relativistic world view; there is no progress without the risk of error.

TWO

VARIATIONS

There are literally dozens of variations on the four-color conjecture that shift its formulation from vertex coloring to edge coloring, touring the vertices of the map, solving diophantine equations, finding the roots of polynomials, and so on. These variations are a testimony to man's tremendous breadth and imagination in his assaults on a difficult problem.

An approach which does not yield a solution after many attempts may need to be modified or abandoned, but this is not proof that it cannot work. It may be that something is missing in the way it was used. If the four-color problem has been solved, we know that all these equivalent formulations are *true*. In other words, the many imaginative equivalent formulations in other apparently far-fetched domains were correctly conceived. The intuition and reasoning involved were right. Of course, shifting the statement of a problem from one area to another may not shed light on its difficulties. In fact, the new area may also be a new field with which we have little familiarity. So it turns out, as we have seen, that the problem was solved by putting it in an equivalent canonical form to study the existence of unavoidable sets of reducible configurations by the method of distributing charges. This, of course, does not mean that there may not be many other ways to solve the problem through another one of the equivalent forms. It is almost certain now (we know it for a fact) that people will continue trying maybe even harder to get a nice elegant solution, since the doubt that the conjecture may not be true has been removed. People would pursue the subject at greater depth and begin to suspect things as being true more readily. In addition, an exhaustive case-by-case solution encourages other exhaustive-type solutions with fewer cases to consider—maybe with proofs that there is a minimum number of cases.

Thus, for those interested, there is even a greater need to know other celebrated equivalent formulations of the 4CC. Our account would not be complete without discussion of those equivalent forms. For most of them, the ideas and proofs of equivalence to the 4CC are basically simple (although their totality may appear tedious) and they are included here. Experience indicates that so far as variety of apparently unrelated ideas are concerned, this problem has probably brought out greater diversity of imagination than any other problem in mathematics. It is to this *monumental* effort spread over a century and to the search for a neat short proof that the rest of the book is dedicated.

FOUR

EDGE COLORINGS:
THE HAMILTONIAN CONNECTION

4-1 INTRODUCTION

By refering to the chart of the introductory chapter (Fig. 1-4), we see that the material of this chapter is concerned with relations depicted in the southwestern corner. They are attempts to prove the 4CC through bridgeless cubic map equivalents. Another variation related to this subject is touring the vertices by a simple circuit (a hamiltonian circuit) which passes by each vertex of a planar graph once and only once. Many of these ideas date back to the early career of the 4CC.

In Sec. 4-2, we introduce the notion of coloring the edges of a cubic map and state an important form of the 4CC due to Tait. We also give an interesting generalization of the 4CC due to Tutte. The next section discusses line graphs. Then, in Sec. 4-4, we study the consequences of hamiltonian circuits and give important conjectures (due to Petersen and Whitney) dealing with the colorability of hamiltonian graphs and maps. Finally, the problem of partitioning the edge set of a graph (to satisfy various conditions) is considered in Sec. 4-5, where we present some slightly esoteric formulations due to Ore, Bondy, and Walther.

4-2 EDGE COLORING IN CUBIC MAPS

A (proper) three-coloring of the edges of a cubic map (called a *Tait-coloring* or an *edge-coloring*) is a coloring of the edges such that any two adjacent edges have different colors, which is the same as saying that the three edges meeting at each vertex have different colors. The following conjecture is due to Tait (1880).

Conjecture C_4 The edges of a bridgeless cubic planar map are three-colorable.

This conjecture will be very important in later chapters since many of the equivalent forms of the 4CC are related to it.

Figure 4-1 shows that the hypothesis "bridgeless" in Conjecture C_4 is necessary since this cubic map cannot be Tait-colored.

The hypothesis of planarity also cannot be omitted because of the Petersen graph P (see Fig. 4-2), which turns out to be the sole smallest-girth graph that has no Tait-coloring. As we shall see a little later from Petersen's conjecture, the shortest-length circuit in P has five edges. But P needs an even-length circuit which here must have length 10 and which includes every vertex once and only once (a hamiltonian circuit) in order to be Tait-colorable. It is easy to show that P has no such circuit. See Tutte (1946) and Isaacs (1975). The latter also gives a generalization of examples by Blanuša (1946), Descartes (1974b), and Szekeres (1973) to infinite families that are not Tait-colorable.

In 1948 Descartes suggested that a fruitful approach to solving the 4CC might lie in a classification of those bridgeless cubic graphs with no Tait-coloring and in 1966 Tutte developed this idea further with a conjecture.

Conjecture T_1 If G is a bridgeless cubic graph with no Tait-coloring, then G contains a subgraph H which can be contracted onto the Petersen graph.

This conjecture needs some explanation. For clarity, we shall assume for the remainder of this section that all graphs are simple. Since the underlying graph of any standard map is simple, there is no harm in this simplification. A *contraction* from a (simple) graph H onto a (simple) graph K is a function $\varphi : V(H) \to V(K)$ from the vertices of H onto those of K satisfying the following conditions:

(1) If $v \in V(K)$ and $\varphi^{-1}(v) = \{w \in V(H) \,|\, \varphi(w) = v\}$, then the full subgraph of H induced by $\varphi^{-1}(v)$ is connected.
(2) If $[v, w] \in E(H)$, then either $\varphi(v) = \varphi(w)$ or $\varphi(v)$ is adjacent to $\varphi(w)$.

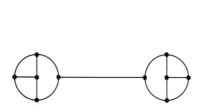

Figure 4-1

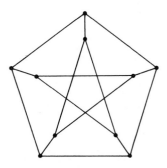

Figure 4-2

(3) If $[x, y] \in E(K)$, then there is an edge $[v, w] \in E(H)$ with $\varphi(v) = x$ and $\varphi(w) = y$.

We also call K a *contraction* of H. An *elementary contraction* of a graph H is obtained by removing two adjacent vertices v and w and replacing them with a new vertex x which is adjacent to all vertices to which either v or w was adjacent. It is not difficult to obtain the following characterization.

Lemma 4-1 K is a contraction of H if and only if K can be obtained from H by a sequence of elementary contractions.

Note that if H is planar and H' is an elementary contraction of H, H' is planar. Hence, we have the following theorem.

Theorem 4-1 Any contraction of a planar graph is planar.

Corollary 4-1 The Petersen graph is not planar.

PROOF Since P can be contracted to K_5, P is not planar.

Now we can explain the significance of Tutte's conjecture.

Theorem 4-2 Tutte's conjecture implies Conjecture C_4.

PROOF Let G be any (simple) cubic bridgeless graph which cannot be Tait-colored. By Tutte's conjecture, G contains a subgraph H which can be contracted to P. By Theorem 4-1 and Corollary 4-1 H is not planar and, hence, neither is G. Thus, any planar bridgeless cubic map can be Tait-colored.

Conjecture C_4' The edges of a triangular map can be colored with three colors so that the edges bounding every triangle are colored distinctly.

This conjecture is simply the dual to Conjecture C_4. The following theorem shows that Conjecture C_2' (in Sec. 3-2) and Conjecture C_4 are equivalent.

Theorem 4-3 A bridgeless cubic map M can be four-colored if and only if its edges can be three-colored.

PROOF Suppose that we are given a bridgeless cubic map M whose regions have been four-colored using colors 0, 1, 2, and 3. We may then Tait-color the edges according to the scheme shown in Table 4-1.

This scheme works because if v is any vertex of M then v lies on the boundary of three distinct regions R, S, and T or else M would have a bridge. Since each pair of the three regions has for a common boundary edge one of the three edges at v, these three edges are all assigned a different color.

Table 4-1

Color edge	If edge lies on boundaries of regions colored
α	0 and 1, or 2 and 3
β	0 and 2, or 1 and 3
γ	1 and 2, or 0 and 3

Conversely, suppose we are given a Tait-coloring of the edges of M using the colors α, β, and γ. Those edges labeled α and β form disjoint circuits (of even length) which we call α–β circuits.

Now every region R of M is contained in the interiors of either an odd or an even number of α–β circuits. Let us pre-color R with $1'$ if R is contained in an odd number of α–β circuits and $0'$ if R is contained in an even number of α–β circuits. Similarly, every region R of M is contained in either an even or odd number of α–γ circuits. In the former case, we pre-color R with $0''$ and in the latter case with $2''$. Now color the regions of M according to the scheme shown in Table 4-2. Thus, each region is pre-colored twice and two regions are colored the same if and only if both of their pre-colorings are the same.

This yields a proper coloring of the regions, for if two regions R_1 and R_2 have a common edge e, then e may be colored either α, β, or γ. If e is colored β, then e lies on exactly one α–β circuit C which contains either R_1 or R_2, but not both, in its interior. Hence, R_1 and R_2 are pre-colored with $1'$ and $0'$ or $0'$ and $1'$, respectively. Thus, they cannot be colored the same. The same argument holds when e is colored γ. If e is colored α, then e lies on both an α–β and an α–γ circuit; so the argument above shows that both pre-colorings of R_1 and R_2 are different and we may again conclude that R_1 and R_2 are colored differently.

The edges of a map are said to be *k-colored* if each edge is assigned one of k colors in such a way that no two adjacent edges have the same color.

Table 4-2

Color region	If region has already been pre-colored
0	$0'$ and $0''$
1	$1'$ and $0''$
2	$0'$ and $2''$
3	$1'$ and $2''$

In view of Conjecture C_4 and Theorem 4-3, it is nice to know that k need not exceed 4 when the map is cubic. The proof of the following result is due to Golovina and Yaglom (1963).

Theorem 4-4 The edges of a cubic map can be four-colored.

PROOF Since a cubic graph is regular of degree 3 it has an even number of vertices, say $2n$. The proof will be by induction on n. When $n = 2$ (that is, when the cubic graph has four vertices) the edges can be easily colored with four colors, say C_1, C_2, C_3, and C_4. Assume that the edges of a cubic graph with $2n$ vertices can be properly colored using only four colors.

Consider, now, a cubic graph G with $2(n + 1)$ vertices. Without loss of generality, remove the edge $[v_i, v_j]$ from G. After removal of this edge, both v_i and v_j will have degree 2. If v_a and v_b are the other two vertices adjacent to v_i, amalgamate the two edges $[v_a, v_i]$ and $[v_i, v_b]$ into a single edge $[v_a, v_b]$ by suppressing vertex v_i. Similarly, if v_d and v_f are the vertices adjacent to v_j, amalgamate the edges $[v_d, v_j]$ and $[v_j, v_f]$ into $[v_d, v_f]$ by suppressing v_j. The resulting cubic graph G' has $2n$ vertices and therefore, by hypothesis, the edges of G' can be four-colored. Then restore the edge $[v_i, v_j]$ such that the vertices v_i and v_j subdivide the edges $[v_a, v_b]$ and $[v_d, v_f]$, respectively. Let (e_a, e_a'), (e_b, e_b'), (e_d, e_d'), and (e_f, e_f') denote the other pairs of edges incident with v_a, v_b, v_d, and v_f, respectively. See Figs. 4-3, 4-4, 4-5, and 4-6 for the sketch of the argument.

Case (1) (See Fig. 4-3)
If $[v_a, v_b] = [v_d, v_f]$, that is, if both v_i and v_j are restored on the same edge of G', then the new edges will be $[v_a, v_i]$, $[v_j, v_b]$, $[v_i, v_j]$, and $[v_i, v_j]'$ where the last two edges are parallel. It is shown in tabular form in Table 4-3 that in this case the four-coloring of graph G' can be extended to the graph G.

Case (2) (See Fig. 4-4)
If $v_a = v_d$ but $v_b \neq v_f$, then the new edges will be $[v_a, v_i]$, $[v_a, v_j]$, $[v_i, v_b]$, $[v_j, v_f]$, and $[v_i, v_j]$. It is shown in Table 4-4 that also in this case the given four-coloring of the edges of G' can be extended to the graph G.

Case (3) (See Fig. 4-5)
If $[v_i, v_j]$ is restored between two parallel edges, then the new edges will be

Figure 4-3

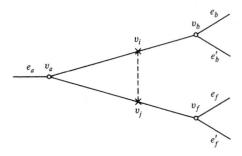

Figure 4-4

Table 4-3

If edges are colored as follows		Then color new edges with colors given below			
e_a, e'_a	e_b, e'_b	$[v_a, v_i]$	$[v_j, v_b]$	$[v_i, v_j]$	$[v_i, v_j]'$
C_1, C_2	C_1, C_2	C_3	C_3	C_1	C_4
C_1, C_2	C_1, C_3	C_4	C_2	C_1	C_3

Table 4-4

If edges are colored as follows			Then color new edges with colors given below				
e_a	e_b, e'_b	e_f, e'_f	$[v_a, v_i]$	$[v_a, v_j]$	$[v_i, v_b]$	$[v_j, v_f]$	$[v_i, v_j]$
C_1	C_1, C_2	C_1, C_2	C_3	C_4	C_4	C_3	C_1
C_1	C_1, C_2	C_1, C_3	C_4	C_3	C_3	C_4	C_2
C_1	C_1, C_2	C_3, C_4	C_4	C_3	C_3	C_4	C_2
C_4	C_1, C_2	C_1, C_3	C_2	C_3	C_3	C_2	C_1

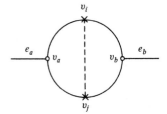

Figure 4-5

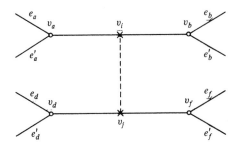

Figure 4-6

$[v_a, v_i]$, $[v_i, v_b]$, $[v_a, v_j]$, $[v_j, v_b]$, and $[v_i, v_j]$. It is shown in tabular form in Table 4-5 that four-coloring the edges of G' can be extended to graph G.

Case (4) (See Fig. 4-6)
If v_a, v_b, v_d, and v_f are all distinct, then the new edges will be $[v_a, v_i]$, $[v_i, v_b]$, $[v_d, v_j]$, $[v_j, v_f]$, and $[v_i, v_j]$. In this case, it is shown in Table 4-6 that for all proper combinations of colorings, except one, of e_a, e'_a, e_b, e'_b, e_d, e'_d, e_f, and e'_f, the four-coloring of G' can be extended to G.

The exceptional combination of colors occurs when the edges e_a and e'_a are colored by C_1 and C_2 and the edges e_b and e'_b are also colored by C_1 and C_2 while the other two pairs of edges are both colored C_3 and C_4. In that case if the color of e_a is C_1, choose the longest trail in G' originating from vertex V_a (with e_a as the first edge) and which consists of edges alternately colored C_1 and C_3 (such a trail may consist of only one edge or it may end at any one of the vertices v_b, v_d, or v_f). Since G' is four-colored, such a trail must be a path. Now interchange the colors of the edges along that path. Obviously G' is still four-colored, but the trouble-making combination of colors at the vertices v_a, b_b, v_d, and v_f is converted into one of the nice forms in Table 4-6 and, therefore, we can now extend the four-coloring of the edges of G' to the graph G.

This completes the proof.

Table 4-5

If edges are colored as follows		Then color new edges with colors given below				
e_a	e_b	$[v_a, v_i]$	$[v_i, v_b]$	$[v_a, v_j]$	$[v_j, v_b]$	$[v_i, v_j]$
C_1	C_1	C_2	C_3	C_3	C_2	C_1
C_1	C_2	C_2	C_3	C_3	C_1	C_4

Table 4-6

If edges are colored as follows				Then color new edges with colors given below				
e_a, e'_a	e_b, e'_b	e_d, e'_d	e_f, e'_f	$[v_a, v_i]$	$[v_i, v_b]$	$[v_d, v_j]$	$[v_j, v_f]$	$[v_i, v_j]$
C_1, C_2	C_1, C_2	C_1, C_2	C_1, C_2	C_3	C_4	C_3	C_4	C_1
C_1, C_2	C_1, C_2	C_1, C_2	C_1, C_3	C_3	C_4	C_3	C_2	C_1
C_1, C_2	C_1, C_2	C_1, C_3	C_1, C_4	C_3	C_4	C_4	C_3	C_1
C_1, C_2	C_1, C_2	C_1, C_3	C_3, C_4	C_3	C_4	C_4	C_2	C_1
C_1, C_2	C_1, C_3	C_1, C_2	C_1, C_3	C_3	C_2	C_3	C_2	C_1
C_1, C_2	C_1, C_3	C_1, C_3	C_1, C_4	C_3	C_2	C_4	C_3	C_1
C_1, C_2	C_1, C_3	C_1, C_3	C_3, C_4	C_3	C_2	C_4	C_2	C_1
C_1, C_3	C_1, C_4	C_1, C_3	C_1, C_4	C_2	C_3	C_2	C_3	C_1
C_1, C_3	C_1, C_4	C_1, C_3	C_3, C_4	C_2	C_3	C_4	C_2	C_1
C_1, C_3	C_3, C_4	C_1, C_3	C_3, C_4	C_4	C_2	C_4	C_2	C_1
C_1, C_2	C_1, C_2	C_3, C_4	C_3, C_4	See the text of proof for this case.				

This result can also be obtained as a corollary to the theorems of Shannon, Vizing, or Brooks (see Chap. 6).

4-3 LINE GRAPHS

The *line* or *interchange graph* $L(G)$ of a graph G is obtained by associating a vertex with each edge of the graph and connecting two vertices by an edge if and only if the corresponding edges of the given graph meet at one or both endpoints (see Fig. 4-7). Clearly a k-coloring of $L(G)$ is just a k-edge-coloring of G. H is a *line graph* if $H \doteq L(G)$ for some graph G.

Conjecture C_5 The vertices of the line graph of a bridgeless cubic planar map can be colored with three colors.

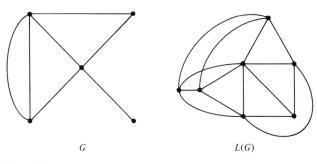

G $L(G)$

Figure 4-7

The equivalence of Conjectures C_4 and C_5 is clear.

We can use line graphs to set up a more exotic reformulation of the 4CC (see Ore, 1967, p. 126). Call a map M *triangle-colored* if every vertex of M has even degree and the regions of M are two-colored using two colors α and β in such a way that all of the α faces are triangles.

Conjecture C_6 The vertices of a planar triangle-colored map which is regular of degree 4 can be three-colored.

Theorem 4-5 Conjectures C_4 and C_6 are equivalent.

PROOF Let M be a map and let $G = U(M)$. Let us modify the thickened map $\overline{M}$ of M (see Fig. 1-12) as follows: for every edge e in M, shrink the dual (orthogonal) edge $\bar{e}$ in $\overline{M}$. Call the new map $\tilde{M}$ (see Fig. 4-8).

If M is cubic then $U(\tilde{M}) = L(G)$. This construction of $L(G)$ when G is planar and cubic is similar to the construction of medial graphs given by Ore (1967, pp. 47 and 124).

Since G is regular of degree 3, $L(G)$ is regular of degree $2(3 - 1) = 4$. We can triangle-color $\tilde{M}$ by coloring every region corresponding to a vertex of

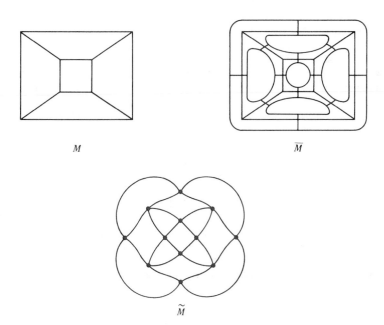

M $\overline{M}$

$\tilde{M}$

Figure 4-8

M with the color α. Note that every "vertex-region" is a triangle while every "region-region" has as many sides as the region of M to which it corresponds. By C_6, we can three-color the vertices of $\tilde{M}$. But this is equivalent to Tait-coloring the edges of M.

Conversely, suppose M is a map regular of degree 4 which is triangle-colored. If $H = U(M)$, it suffices to show that $H = L(G)$, where G is planar cubic and bridgeless, for then, by C_4, G can be Tait-colored, and hence $H = L(G)$ can be three-colored. We construct G as follows. Let $\mathscr{R}$ be the set of triangular regions colored α in the map M. Put a vertex v_R in every region $R \in \mathscr{R}$ and join two vertices v_R and v_S by one edge for every corner which the corresponding regions have in common. Since $U(\tilde{M})$ is simple, so is G. Moreover, G is cubic, bridgeless, and planar, and $L(G) = H$.

4-4 THE HAMILTONIAN CONNECTION

In this section, we introduce the very important idea of hamiltonian graphs. Because of the extreme utility of this concept in all areas of graph theory, we shall dwell on it at some length.

A graph is said to be *hamiltonian* if it has a simple circuit called a *hamiltonian* circuit which passes through each vertex exactly once. The next result relates hamiltonian circuits to map-coloring.

Theorem 4-6 Let M be a cubic map such that $U(M)$ is hamiltonian. Then $\chi(M) \le 4$.

PROOF Let C be a hamiltonian circuit in $U(M)$. Since M is cubic, C has an even number of vertices, and, hence, an even number of edges. If two colors are alternately assigned to the edges of C and the third color assigned to the remaining edges, we have Tait-colored M, and this implies that M is four-colorable. Hence, every cubic hamiltonian map can be four-colored.

The corresponding statement for planar hamiltonian graphs was shown by Whitney (1931) to be equivalent to the 4CC.

Conjecture C_7 Every hamiltonian planar graph is four-colorable.

Before proving the equivalence of Conjectures C_7 and C_1, we need another result due to Whitney (1931). A *separating triangle* is a circuit of length 3 whose deletion disconnects the graph. The nonexistence of separating triangles in a maximal planar graph G is equivalent to G being four-connected when G has at least five vertices (see Appendix).

Theorem 4-7 Let G be a maximal planar graph without separating triangles. Then G is hamiltonian.

The proof of the theorem consists of a lengthy enumeration of cases and we shall not give it here.

Theorem 4-8 Conjectures C_1 and C_7 are equivalent.

PROOF It is clear that C_1 implies C_7. Conversely, suppose that C_1 is false. Then there exists a minimal five-chromatic planar graph G which is necessarily maximal planar. We claim that G is also a counter-example to C_7. Suppose that G contains a separating triangle T. Then $G-T$ is a disconnected graph with two components G_1 and G_2. Let $\overline{G}_i = G_i \cup T$ $(i = 1, 2)$. Then $\overline{G}_1$ and $\overline{G}_2$ can be four-colored. But $\overline{G}_1 \cap \overline{G}_2 = T$, and it is plain that we can permute the colors in a four-coloring of $\overline{G}_2$ to make it agree with the four-coloring of $\overline{G}_1$ on T. This yields a four-coloring of G, which is impossible.

Conjecture C_8 It is possible to four-color the vertices of a planar graph consisting of a regular polygon of n sides with noncrossing diagonals dividing the interior and exterior of the polygon into triangular regions.

Conjecture C_8 is equivalent to Conjecture C_7, since a maximal planar hamiltonian graph is just a polygon with noncrossing diagonals dividing its interior and exterior regions into triangles.

The following conjecture, due to Petersen (1891) and Tait (1880a), relates the 4CC to hamiltonian circuits and the traversality of certain graphs.

Conjecture C_9 In a bridgeless cubic map, it is possible to either tour all the vertices by a hamiltonian circuit or to find a group of disjoint even-length circuits covering all the vertices.

Theorem 4-9 Conjectures C_9 and C_4 are equivalent.

In fact, the equivalence really has nothing to do with maps or planarity, so we prove the following theorem instead.

Theorem 4-10 Let G be a cubic graph. Then the edges of G can be three-colored if and only if there exists a collection of disjoint even-length cycles covering the vertices of G.

PROOF Assume first that the edges of G are three-colored with α, β, and γ. Let $G_{\alpha\beta}$ denote the full subgraph of G induced by the set of edges colored either α or β. Then $G_{\alpha\beta}$ is a spanning subgraph of G and is regular of degree 2. Now, any graph regular of degree 2 is a disjoint union of cycles, so we need only note that each cycle of $G_{\alpha\beta}$ has even length since the edges of $G_{\alpha\beta}$ can be two-colored.

Conversely, given the collection of disjoint even-length cycles covering the vertices of G, two-color the edges in each cycle using α and β and assign γ to the remaining edges. This is a three-coloring of the edges of G.

In view of Conjectures C_7 and C_9, we might begin to wonder when an arbitrary connected graph is hamiltonian. Sufficient conditions for a graph to be hamiltonian are found in the Appendix. Note that Theorem 4-10 implies that if a bridgeless cubic map has an underlying graph which is hamiltonian, then the map can be four-colored.

Tait (1884) once conjectured that every three-connected cubic map is hamiltonian, but Tutte (1946) gave a counter-example with 46 vertices (see Fig. 4-9). Had Tait's conjecture been true, the truth of Conjecture C_0 would have followed, for, in order to prove Conjecture C_0, it suffices to show that every standardized map M can be four-colored. But any standardized map M can be regarded as a flattening out in the plane of a convex three-dimensional polyhedron. Steinitz's theorem (see Appendix) assures us that $U(M)$ is three-connected. Tait's conjecture would then imply that every such map had a hamiltonian circuit and, hence, was four-colorable.

Tait himself did not supply an adequate proof as to how the four-color conjecture would be true if his conjecture were true. He thought his conjecture was true from all the evidence he had. Chuard (1932) went on to "complete" the story. Doubts as to the validity of Chuard's claim were expressed by Pannwitz (1932). In any event, Tutte's example has made the entire debate academic as a means of settling the four-color conjecture.

4-5 EDGE PARTITIONS AND FACTORIZATION

A graph G is *k-factorable* if it can be written as the edge-disjoint union of subgraphs G_i in such a way that every vertex $v \in V(G)$ meets exactly k edges in each G_i. The subgraphs G_i are called *k-factors* of G.

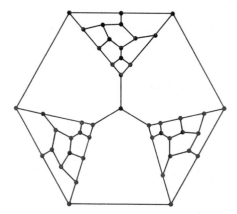

Figure 4-9

Conjecture C_{10} Every cubic bridgeless planar map is one-factorable.

This conjecture is obviously equivalent to Conjecture C_4 since a one-factorization is equivalent to an edge-coloring. An interesting variant on this conjecture is given in Ore (1967, p. 103).

Conjecture C_{11} Every bridgeless planar graph is the union of three edge-disjoint subgraphs such that each vertex has either an even number of edges incident with it from each of the three subgraphs or it has an odd number from each of them.

Plainly, Conjecture C_{11} implies C_4 since if $3 = p_1 + p_2 + p_3$, where each p_i is a nonnegative integer and all the p_i have the same parity, then $p_i = 1$ for $i = 1, 2, 3$. Hence, a cubic graph satisfying the conclusion of C_{11} is Tait-colored.

Theorem 4-11 Conjecture C_{11} is equivalent to Conjecture C_0.

PROOF We have already seen that C_{11} implies C_4 and hence C_0. Conversely, suppose C_0 holds and let G be any bridgeless planar graph. By working with one component at a time, we may assume G connected. Draw G in the plane to produce a map M with $U(M) = G$. Let $T(M)$ be the cubic map obtained by "inflating" every vertex (see Fig. 1-12) and let $G' = U(T(M))$. Since M is bridgeless, $T(M)$ is bridgeless. By Conjecture C_0 we can four-color the regions of $T(M)$ using colors 0, 1, 2, and 3. Now $T(M)$ is bridgeless so, as in the proof of Theorem 4-3, we can produce a Tait-coloring of $T(M)$ by coloring an edge α if it separates regions colored $(0,1)$ or $(2,3)$; β if it separates $(0,2)$ or $(1,3)$; and γ if it separates $(1,2)$ or $(0,3)$.

Clearly, there is a contraction from G' onto G such that each edge e of G is the image of a unique edge e' of G'. Label e with the color of e'. Let G_α be the subgraph of G determined by all edges labeled α, G_β by those labeled β, and G_γ by those labeled γ. Certainly, G is the edge-disjoint union of G_α, G_β, and G_γ.

It remains to show that each vertex is incident with either an even number of α, of β, and of γ edges or with an odd number of α, of β, and of γ edges. Equivalently, we must show that if $\{x, y\}$ is any two-element subset of $\{\alpha, \beta, \gamma\}$, then each vertex is incident with an even number of edges labeled either x or y. Suppose $\{x, y\} = \{\alpha, \beta\}$ and consider a vertex v in G. If the region $T(v)$ in $T(M)$ corresponding to v was colored 0, then the edges labeled α and β at v correspond to adjacencies in $T(M)$ between regions adjacent to $T(v)$ and colored $(2,3)$ or $(1,3)$, respectively. Therefore, each region of $T(M)$ adjacent to $T(v)$ and colored 3 produced exactly two edges incident with v and labeled α or β, and this is the only way such edges are produced. Thus, the number of such edges at v is even if $T(v)$ is colored 0. All the other cases are identical to this one.

This completes the proof.

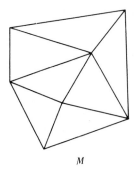

M   **Figure 4-10** *M*.

Note that in forming *T(M)*, we do not really need to inflate vertices of degree 3. Figures 4-10 to 4-14 illustrate the above proof.

The next result is a slight variation on Conjecture C_{10}.

Theorem 4-12 The edges of a bridgeless cubic graph can be colored with two colors α and β so that each vertex is incident with one edge colored with α and two edges colored with β.

This theorem is due to Petersen (1891). It can be restated in the form: every bridgeless cubic graph is the union of a one-factor and a two-factor. The Petersen graph (see Fig. 4-2) shows that a similar result with three one-factors cannot be obtained. A cubic graph with bridges may not have a one-factor (see Fig. 4-15).

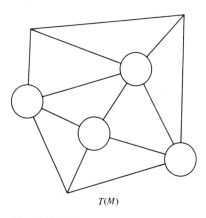

T(M)

Figure 4-11 *T(M)*.

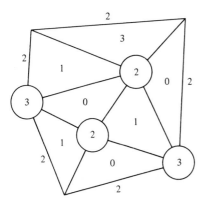

Figure 4-12 $T(M)$ with faces four-colored using the colors 0, 1, 2, and 3.

Another beautiful theorem on the existence of a one-factor is due to Tutte (1947). A component of a graph G is called *odd* if it has an odd number of vertices. Let $k_0(G)$ denote the number of odd components.

Theorem 4-13 A graph G has a one-factor if and only if, for any subset $X \subset V(G), |X| \geq k_0(G-X)$.

PROOF Note that Tutte's theorem actually implies Petersen's, as the following argument (see Berge, 1973, pp. 160–161) shows.

Suppose $X \subset V(G)$ is any nonempty subset with $X \neq V(G)$, where G is any connected cubic bridgeless graph. Let C_i denote any odd component of

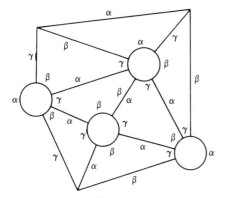

Figure 4-13 $T(M)$ with edges colored α, β, and γ using the scheme as stated in the proof.

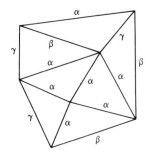

Figure 4-14 M, after contracting the inflated vertices of $T(M)$ and labeling the edges as in $T(M)$.

$G{-}X$, $1 \le i \le k_0(G{-}X)$ and let $m_i = m(X, C_i)$ be the number of edges in G joining vertices of X and of C_i. Since G is connected, $m_i \ge 1$. If $m_i = 1$, G would contain a bridge. If $m_i = 2$, then consider the sum of the C_i degrees:

$$\sum_{v \in V(C_i)} \deg(v) = 3 \left| V(C_i) \right| - 2$$

Since $\left| V(C_i) \right|$ is odd, this sum is odd. But, by the first theorem in graph theory, $\sum_{v \in V(C_i)} \deg(v) = 2q_i$, where q_i is the number of edges in C_i. Thus, $m_i \ge 3$.

Therefore, for any proper nontrivial subset X of $V(G)$, if $m(X, V(G) - X)$ denotes the number of edges incident to only one vertex of X, we have

$$3 \left| X \right| \ge m(X, V(G) - X) \ge \sum_{i=1}^{k_0(G{-}X)} m_i \ge 3k_0(G{-}X)$$

so $\left| X \right| \ge k_0(G{-}X)$. The cases $X = \varnothing$ or $X = V(G)$ are trivially valid and, hence, by Tutte's theorem, G has a one-factor.

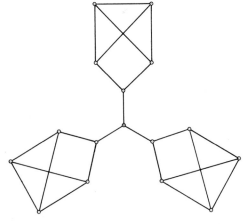

Figure 4-15

Another consequence of Theorem 4-13 is the König–Hall theorem (see Hall, 1934).

Theorem 4-14 A bipartite graph G with color classes X and Y has a one-factor if and only if $|X| = |Y|$ and $|N(S)| \geq |S|$ for all $S \subset X$, where $N(S) = \{w \mid w \sim v$ for some $v \in S\}$.

An interesting variant of Conjectures C_9 and C_{10} is due to Bondy (1974). If G is any graph, a *two-labeling* of G is a partition $C = (B, W)$ of $V(G)$. Let $v(S)$ denote the number of edges of G with exactly one endpoint in S. C is *balanced* if for all $S \subset V(G)$, $v(S) \geq w(S, C)$, where $w(x, C) = -2$ if $x \in B$ and $+2$ if $x \in W$, and $w(S, C) = \sum_{x \in S} w(x, C)$. C is *equitable* if $|B| = |W|$.

Lemma 4-2 If C is balanced, then C is equitable.

Lemma 4-3 If C is balanced and G is cubic, then there is no monochromatic (all B or all W) path of three vertices.

The proofs of these lemmas are exercises.

Conjecture C_{12} Every cubic bridgeless planar graph G has a balanced labeling.

Theorem 4-15 Conjectures C_4 and C_{12} are equivalent.

Before proving this result, we need a few easy facts about balanced colorings. The proofs are left as exercises.

Lemma 4-4 Any even-length cycle has a balanced labeling.

Lemma 4-5 If each component of G has a balanced labeling, so does G.

Lemma 4-6 If G' is a spanning subgraph of G and G' has a balanced labeling, so does G.

PROOF OF THEOREM 4-15 Using Lemmas 4-4, 4-5, and 4-6, it is trivial to show that C_9 implies C_{12}. Conversely, suppose C_{12} holds and let $C = (B, W)$ be a balanced two-labeling of G. If $E' = \{e = [v, w] \in E(G) \mid w(\{v, w\}, C) = 0\}$ and $G' = G(E')$, the spanning subgraph of G determined by E', then G' is bipartite with color classes B and W. Let $V'' = \{v \in V(G) \mid \deg_{G'}(v) = 3\}$, where $\deg_{G'}(v)$ means degree in G', and let $G'' = G(V'')$.

We claim that G'' has a one-factor, E''. Accepting the claim for a moment, we consider $G' - E''$, which is a bipartite spanning subgraph of G regular of degree 2 so that C_9 is satisfied.

To get the one-factor E'', we apply the König–Hall theorem. Suppose G'' has no one-factor. Without loss of generality, choose $P \subset B$ and $Q \subset W$ such that $|P| > |Q|$ and $N''(P) \subset Q$, where $N''(P) = \{w \in V(G'') \,|\, w \sim v$ for some $v \in P\}$. Let R be the set of W-vertices not in $V(G'')$. Then $S = P \cup Q \cup R$ violates the condition for balanced labelings, for if $|P| = p$, $|Q| = q$, and $|R| = r$, then $w(S, C) = 2(q + r - p)$. But $N(P) \subset Q \cup R$, so $v(S) \leq 3q + 2r - 3p < 2(q + r - p) = w(S, C)$.

If we generalize the notion of factorization slightly, we get the concepts of *bipartite dichotomy* (Ore, 1967, p. 104) and *bipartite thickness* (due to Walther). A *bipartite dichotomy* in a graph G is a partition $V(G) = V_1 \cup V_2$ of the set of vertices such that both of the induced subgraphs $G(V_1)$ and $G(V_2)$ are bipartite. The *bipartite thickness* $b\theta(G)$ of a graph G is the minimum number r of bipartite subgraphs $B_1, \ldots, B_r$ of G needed in order that $G = B_1 \cup \cdots \cup B_r$.

The following theorem is not difficult to prove.

Theorem 4-16 For any graph G, the following are equivalent:
(a) $\chi(G) \leq 4$.
(b) G has a bipartite dichotomy.
(c) $b\theta(G) \leq 2$.

ALGEBRAIC AND ARITHMETIC VARIATIONS

5-1 INTRODUCTION

Combinatorial problems come in many guises. For example, the 4CC can be reformulated as a problem on congruences and has other numerical and number theoretic aspects. In this chapter, we shall use congruences and modular arithmetic, together with Galois fields and Diophantine inequalities, arrangements and sequences, and the properties of directed graphs, to examine the 4CC in some purely combinatorial forms.

Heawood attempted very early to translate the 4CC into a more algebraic question. His conjecture and an interesting variation by Hadwiger (the "lesser" Hadwiger conjecture) are given in Sec. 5-2. Galois fields are used to formulate the approach of Sec. 5-3. In Sec. 5-4, we skip ahead to modern times with two intriguing statements due to Gomory and Dantzig. Then Sec. 5-5 brings us to the first completely algebraic formulations of the problem, due to Whitney and later Tutte (as Descartes). More recent versions by Mycielski and Bernhart are also presented. In the last section, we cover Minty's flow-ratio interpretation, as well as a nice approach to coloring due to Gallai.

5-2 VERTEX CHARACTERS

We have already seen several instances, such as the proof that Tutte's theorem implies Petersen's in Sec. 4-5, in which all we needed to know about two numbers is whether or not they had the same parity. More generally, call two integers x and y *congruent modulo k*, written

$$x \equiv y \pmod{k}$$

if x-y is divisible by k.

Heawood (1898) related the 4CC to the solution of a system of congruences modulo 3.

Conjecture C_{13} It is possible to associate a coefficient $k(v)$ equal to $+1$ or -1 with each vertex in a bridgeless cubic map in such a way that, for each region R, $\Sigma k(v) \equiv 0 \pmod{3}$, where the summation is taken over the vertices occurring in the boundary of R.

Such a function k is called a *Heawood vertex character*. A reformulation of this conjecture would be to take the above congruences and require a solution for all of them taken together, none of whose members is congruent to zero modulo 3.

To see how this conjecture implies Conjecture C_4, label the edges of the map a, b, or c such that the three edges incident with each vertex are labeled differently and the ordering of the edges $a \to b \to c$ is a clockwise rotation if $k(v) = +1$ and counterclockwise if $k(v) = -1$. This labeling is consistent if and only if the vertex character assignment is proper; i.e., for each region R, $\Sigma k(v) \equiv 0 \pmod{3}$, where the sum is taken over all vertices in the boundary of R.

On the other hand, if the edges of a bridgeless cubic map have already been Tait-colored a, b, and c, then we define a function k on the vertices by letting $k(v) = +1$ if $a \to b \to c$ is a clockwise rotation and $k(v) = -1$ if counterclockwise. It is easy to check that $\Sigma k(v) \equiv 0 \pmod{3}$ if the sum is taken over all vertices in the boundary of a region.

Thus, we have proved the following theorem.

Theorem 5-1 Conjectures C_4 and C_{13} are equivalent.

In fact, we have actually shown somewhat more—namely, that if M is a map, finding an assignment k as above is equivalent to Tait-coloring the edges.

Corollary 5-1 Let M be a map each of whose regions has a number of sides divisible by 3. Further assume that M is bridgeless and cubic. Then $\chi(M) \leq 4$.

PROOF Assign the coefficient $k(v) = 1$ to every vertex. Then k satisfies the hypothesis of Conjecture C_{13}. Hence, by the preceding remarks, we can Tait-color the edges of M. But this induces a four-coloring of the regions of M.

Conjecture C_{14} It is always possible repeatedly to cut off corners (replace a vertex by a triangle) from a convex (cubic) polyhedron so that eventually a polyhedron is obtained whose faces have a number of edges which is divisible by 3.

This conjecture due to Hadwiger (1957) is essentially a modification of the previous conjecture of Heawood.

Theorem 5-2 Conjectures C_{13} and C_{14} are equivalent.

PROOF First note that Heawood's conjecture for arbitrary cubic bridgeless maps is equivalent to the conjecture for cubic bridgeless maps in which the boundaries of distinct regions intersect in either nothing, a single vertex, or a single edge. (Simply replace M with M_3 and note that a four-coloring of M_3 induces a four-coloring of M. See Sec. 3-2 following Conjecture C_2.) But such maps are convex polyhedra by Steinitz's theorem (see Appendix), since their underlying graphs are clearly three-connected. (In fact, one often *defines* a convex polyhedron to be such a map.) This argument also demonstrates that "convex polyhedron" could be replaced by "bridgeless cubic map" in Conjecture C_{14}.

Consider any cubic polyhedron M. If k is a Heawood vertex character, then, by replacing with a triangle each vertex v for which $k(v) = -1$, we obtain a new polyhedron M' all of whose regions have a multiple of three sides. Conversely, if we can replace certain vertices v by triangles so that the resulting polyhedron has this property, then set $k(v) = -1$ for these vertices and $k(v) = +1$ for all the others we get a vertex assignment (see Fig. 5-1).

To verify the above assertions, let R be any region with t vertices $v_1, \ldots, v_t$ and let k be any function assigning $+1$ or -1 to each vertex. Let $m = |k^{-1}(-1)|$ and $p = |k^{-1}(1)|$. Clearly, $p + m = t$ and $p - m \doteq \sum_{i=1}^{t} k(v_i)$. If each vertex $v_j \in k^{-1}(-1)$ is replaced by a triangle, the resulting region has $t + m$ sides. But $t + m = p + m + m = p + 2m = (p - m) + 3m$ and so $t + m \equiv p - m \pmod{3}$ and hence k is a Heawood vertex character if and only if the process described in C_{14} can be carried out.

Conjecture C_{14} may have been suggested by a result of Heawood (1898) in which he proved that if the regions of a map could each be subdivided (by the simple operation of adding a new edge to connect some pairs of adjacent edges,

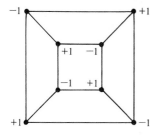

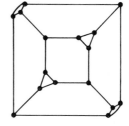

Figure 5-1

thereby forming triangles) into new regions such that all the regions are bordered by edges whose number is congruent to zero (mod 3), then the map is four-colorable. Heawood first shows constructively that such a map is four-colorable. Then he shows that any four-coloring of the constructed map is also a four-coloring of the initial map by removing the edges.

Using a computer code, Yamabe and Pope (1961) have developed an assignment method for cubic maps of up to 36 vertices.

Before continuing with another conjecture of this type, we have to introduce an important class of graphs. A *directed graph* (or *digraph*) consists of a nonempty set V, a set A disjoint from V, and a mapping ∇ of A into $V \times V$. The elements of V and A are, respectively, called *vertices* and *arcs* (or *directed edges*), and ∇ is called the *directed incidence mapping* associated with the directed graph. If $a \in A$ and $\nabla(a) = (v, w)$, then arc a is said to have v as its *initial* vertex and w as its *terminal* vertex. The in-degree (out-degree) of a vertex v is the number of arcs having v for terminal (initial) vertex and is denoted $\mathrm{id}(v)$ ($(\mathrm{od}(v))$.

Directed graphs are usually denoted by D or (V, A, ∇), or by (V, A) when ∇ is not used explicitly. An associated undirected graph is obtained from a directed one by disregarding the ordering of the endpoints of each arc.

By analogy with the undirected case, we have *arc progressions, cycles,* and *paths* instead of edge progressions, circuits, and chains. "Cycles" is sometimes used by authors interchangeably with "circuits." So far as is possible we have attempted to standardize the terminology as indicated here, but the reader should be aware that in the literature cycle is often used instead of circuit.

In Fig. 5-2 below, there are two paths from v to w, one of which consists of the arc a. If the direction of a is reversed, there will then be a cycle. With its present orientation, however, this directed graph contains no cycles. (Such digraphs are called *acyclic.*) The in-degree of v is 0 but $\mathrm{od}(v) = 2$.

Synge (1967) has developed two similar conjectures relating to the "consistent unambiguous" four-coloring of the vertices and three-coloring the arcs of a directed triangular network with $n + 2$ vertices, $3n$ edges, and $2n$ regions. Here is a sketch of the idea. The "coded" colors assigned to the vertices are the 4 fourth roots of unity, $i, i^2, i^3,$ and 1. The regions are oriented with counterclockwise as positive but the outside region (which goes to infinity) is oriented so that the clockwise sense on its inner boundary is positive. An arbitrary positive sense is assigned to each edge. Each edge is then assigned a number j from the set (i, i^2, i^3). If vertices v and v' are adjacent and c, a color from $(i, i^2, i^3, 1)$, is assigned to v, then v' is assigned $c' = jc$ if the sense of the arc is from v to v' and $c' = j^{-1}c$ otherwise. This ensures

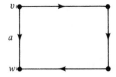

Figure 5-2

different colors for v and v' and makes it possible to recover the color of v by returning from v'. Let j be arbitrarily assigned to all the edges. Start with v and assign it a color, and extend to all other vertices using the above rule for the coloring of adjacent vertices. However, if a circuit is followed, v may not recover its original color. This leads to a set of consistency conditions which must be satisfied.

Since each region is a triangle, let I_1, I_2, and I_3 be the incidence indices of its edges ($= +1$ if the sense of the edge agrees with that of the region and $= -1$ if not). Let j_1, j_2, and j_3 be the color instructions assigned to these edges. If we start with a vertex colored c on the boundary of the region, go around the boundary in a positive sense and return to it, its color is now $c' = wc$, where $w \equiv j_1^{I_1} j_2^{I_2} j_3^{I_3}$ is called the weight of the region and we can only have $w = +1, +i$. The consistency condition for the region is $w = 1$.

Conjecture C'_{14} It is possible to assign color instructions j from the set (i, i^2, i^3) to the edges of a triangular network to make the weights of every region equal to unity.

Synge proves the following theorem.

Theorem 5-3 Conjectures C_1 and C'_{14} are equivalent.

DISCUSSION Since there are $2n$ regions we want $\Sigma w = 2n$, and since w can take on the values $\pm 1, \pm i$, let x regions have $w = 1$, y have $w = -1$, z have $w = i$, and t have $w = -i$. Then $\Sigma w = x - y + i(z - t)$, $x + y + z + t = 2n$ and hence $\Sigma w = 2n$ if and only if $x = 2n$, $y = z = t = 0$. Thus with any choice of color instruction, Σw is a complex number which lies in the square $(\pm 2n, \pm 2ni)$. The object is to change the instructions step by step to reach $\Sigma w = 2n$ from any initial assignment. This is somewhat reminiscent of discharging methods. Now if the real part of $\Sigma w = M < 2n$, then the nonnegative index of uncolorability of a map is given by $2n - M$, whose vanishing is necessary and sufficient for four-colorability.

A systematic way to test the 4CC would be to start with the worst choice $j = i^2 = -1$ for every edge; hence $\Sigma w = -2n$. Number the edges $1, 2, \ldots, 3n$. There are 3^{3n} possible instructions each time a choice is made from (i, i^2, i^3), yielding a point in the complex plane for Σw. If the conjecture is true, $\Sigma w = 2n$ is among these values; otherwise not.

5-3 MODULAR EQUATIONS AND GALOIS FIELDS

A field is an algebraic system consisting of a set (infinite or finite) and two binary operations (addition and multiplication) defined for all pairs of elements such that the set is a nontrivial commutative group under addition and its nonzero elements form another commutative group under multiplication. Furthermore, multiplication must distribute over addition. The best-known example is the set of real numbers. The integers modulo a prime integer p form what is known as a modular field denoted by Z_p—the simplest form of a finite field. If F is any field we may form the algebraic structure (called an integral domain) of all polynomials with coefficients in F, denoted by $F[x]$. Let $p(x)$ be a monic polynomial (a polynomial with leading coefficient 1) which cannot be factored into lower-degree polynomials with coefficients in F. The set of residues of all polynomials in $F[x]$ divided by multiples of $p(x)$—these multiples are denoted by $(p(x))$—forms a field. This field is called a simple algebraic extension of F. Now the characteristic of a field F is the least positive integer k such that $kx = 0$ for any x in F. It turns out that the characteristic of any field is a prime p or infinite. Any finite field of characteristic p is a simple algebraic extension of its prime subfield Z_p. It contains p^n elements for some positive integer n and is given by $Z_p[x]/(p(x))$ (division by $p(x)$ as above) for an irreducible monic polynomial $p(x)$ of degree n. All such fields are isomorphic. We denote this field by $GF(p^n)$ in honor of Evariste Galois who first constructed them.

Let $GF(k)$ denote the Galois field of order k. Thus, k is a prime power and $GF(k)$ is the unique (finite) field with k elements. Obviously, one may view a k-coloring of the vertices (or edges or regions) of a graph (or map) as an assignment of an element of $GF(k)$ to every vertex (or edge or region) of the graph (or map). Let $GF(k)^*$ denote the set of nonzero elements in $GF(k)$. We shall consider in this section the cases $k = 2, 3, 4$.

Note that two elements in $GF(4)$ are equal if and only if their sum is zero, since the characteristic of this field is 2 so that $x + x = 2x = 0$.

If we assign to every edge e in a bridgeless map which has been four-colored the sum of the colors of the two regions adjacent to e, this sum belongs to $GF(4)^*$. Thus, the edges are three-colored since there are three nonzero elements in $GF(4)$.

We may give this a matrix formulation as follows. List the edges $e_1, \ldots, e_m$ and regions $r_1, \ldots, r_n$ of a bridgeless map M. Let B be the matrix defined by putting $B_{ij} = 1$ if e_i is in the boundary of r_j and letting $B_{ij} = 0$ otherwise. Thus, each row of B contains two 1's and all the rest 0's. B is sometimes called the *edge-region incidence matrix* of M, or simply an *incidence matrix*.

Suppose M is four-colored. Then define a column vector $Z = (z_1, \ldots, z_n)$, where z_j is the color of the jth region and each z_j belongs to $GF(4)$. We write $Z \in GF(4)$ for short. The matrix product BZ is a column vector $p = (p_1, \ldots, p_m)$ and each p_i is the sum of two distinct elements in $GF(4)$ since e_i is on the boundary of two distinctly colored regions. Hence, each p_i is nonzero.

Now we can state the following conjecture due to Veblen (1913).

Conjecture C_{15} Let B be any edge-region incidence matrix. Then there is a vector $Z \in GF(4)$ such that the matrix product $BZ \in GF(4)^*$.

The discussion above shows that Conjecture C_{15} is equivalent to Conjecture C_0 since the existence of the column vector Z provides us with a four-coloring of the map.

If we define a region-vertex incidence matrix for a map M in the obvious way, we can then make the following conjecture, which is simply a restatement of Conjecture C_{13} using the Galois field GF(3).

Conjecture C_{16} Let B be the region-vertex incidence matrix of a map M. Then there is a vector $Z \in$ GF(3)* such that $BZ = 0$.

5-4 DIOPHANTINE INEQUALITIES

Let the regions of a planar map M be labeled $r = 1, 2, \ldots, n$. Let the variable t_r be integer-valued $0 \leq t_r \leq 3$. Thus, t_r assigns one of the four colors, labeled 0, 1, 2, 3, to the region whose number is r. If two regions r and s have a boundary in common, then $t_r - t_s \neq 0$. Such a relation is written down for every pair of adjacent regions. The relation for one pair may be reduced to two inequalities as follows:

$$\text{either} \quad (a) \ t_r - t_s \geq 1 \quad \text{or} \quad (b) \ t_s - t_r \geq 1 \tag{5-1}$$

This pair of inequalities can be written as

$$t_r - t_s \geq 1 - 4\delta_{rs} \tag{5-2}$$

and

$$t_s - t_r \geq -3 + 4\delta_{rs}, \quad \text{where} \quad \delta_{rs} = 0 \text{ or } 1 \tag{5-3}$$

We obtain a system $\mathscr{S}(M)$ of such inequalities by allowing r and s to vary from 1 to n over all pairs $r \neq s$ of adjacent regions. The point is that we always prefer to have the *conjunction* of inequalities, rather than the disjunction as a constraint, since this is the suitable form for linear programming or integer programming.

Lemma 5-1 The inequalities (5-1) (a) or (b) are equivalent to the inequalities (5-2) and (5-3).

PROOF If (5-1)(a) holds, set $\delta_{rs} = 0$. Then (5-2) is identical to (5-1)(a) and (5-3) becomes

$$t_s - t_r \geq -3$$

which is trivially true. If (5-1)(b) holds, set $\delta_{rs} = 1$ and check that again (5-2) and (5-3) hold. The converse is equally straightforward.

We have now proved that the following conjecture is equivalent to Conjecture C_0.

Conjecture C_{17} For any planar map M with n regions, it is possible to choose the binary variables δ_{rs} such that the system $\mathscr{S}(M)$ of inequalities has a solution $(t_1, \ldots, t_n)$ where each t_i is an integer $0 \leq t_i \leq 3$.

According to Dantzig, this formulation was informally communicated to him by Ralph Gomory of integer programming fame.

Another formulation of the 4CC is due to Dantzig himself (1963, p. 549). Referring back to Conjecture C_9, consider one of the even-length covering circuits C in a planar bridgeless cubic graph and, starting at any vertex v covered by C, assign a direction to one of the edges e in C incident with v. Now assign the opposite direction to the other edge of C incident with v, and continue around the (even-length) circuit in this manner so that for each vertex the two edges incident with it (now called arcs) are directed away from it or directed towards it.

In the resulting directed graph D, label the vertices $1, 2, 3, \ldots, n$. For any pair of adjacent vertices i and j, we write $x_{ij} = 1$ if there is an arc directed from i to j. Otherwise we write $x_{ij} = 0$.

By construction every vertex v of G has either

$$(a') \quad \text{od}(v) = 2 \quad \text{and} \quad \text{id}(v) = 0$$

or

$$(b') \quad \text{od}(v) = 0 \quad \text{and} \quad \text{id}(v) = 2 \tag{5-4}$$

These translate into inequalities as follows:

$$(a) \quad \sum_j x_{ij} = 2 \quad \text{and} \quad \sum_j x_{ji} = 0$$

or

$$(b) \quad \sum_j x_{ij} = 0 \quad \text{and} \quad \sum_j x_{ji} = 2 \tag{5-5}$$

Just as (5-1) (a) or (b) were rewritten as (5-2) and (5-3), we can rewrite (5-5) (a) or (b) as

$$\sum_j x_{ij} = 2\delta_i \tag{5-6}$$

and

$$\sum_j x_{ji} = 2 - 2\delta_i \tag{5-7}$$

where $\delta_i = 0$ or 1 is an appropriately chosen binary variable.

Let $\mathscr{S}'(G)$ denote the system of these $2n$ inequalities. The preceding discussion shows that Conjecture C_9 implies the following conjecture which reformulates the 4CC as a bounded transportation problem to which the techniques of integer programming may be applied.

Conjecture C_{18} Let G be a bridgeless planar cubic graph with n vertices $1, \ldots, n$. For all i, j, set $y_{ij} = 1$ if i and j are adjacent and $y_{ij} = 0$ otherwise. For $i = 1, \ldots, n$, we can choose binary variables $\delta_i (1 \le i \le n)$ such that the system $\mathscr{S}'(G)$ has a solution $\{x_{ij} \mid 1 \le i, \ j \le n\}$, where the x_{ij} are binary variables satisfying the additional constraint $x_{ij} \le y_{ij}$ for all i, j.

Theorem 5-4 Conjecture C_9 and C_{18} are equivalent.

PROOF We have already seen that C_9 implies C_{18}. Conversely, suppose C_{18} holds and let G be any planar cubic bridgeless graph with vertices $1, \ldots, n$.

By C_{18} we can find binary variables $\delta_i (1 \leq i \leq n)$ and a corresponding solution x_{ij} of $\mathscr{S}'(G)$ satisfying $x_{ij} \leq y_{ij}$ as well. Consider the spanning subgraph H of G in which i and j are adjacent if and only if $x_{ij} + x_{ji} = 1$. Now form a directed graph D from H by directing an edge from i to j if $x_{ij} = 1$ and from j to i if $x_{ji} = 1$. For every vertex v of D, we have $\text{od}(v) = 2$ and $\text{id}(v) = 0$ or $\text{od}(v) = 0$ and $\text{id}(v) = 2$ because of (5-6) and (5-7). Thus, H is regular of degree 2 and so decomposes as the disjoint union of circuits, and, by the condition on D, each of these circuits must have even length. But this is precisely what is asserted by C_9.

5-5 ARRANGEMENTS AND SEQUENCES

Whitney (1937) inaugurated the technique of studying the 4CC as a numerical problem. We shall give two of his conjectures and show their equivalence to the 4CC. While the conjectures appear to be purely number-theoretic, there is geometry lurking in the algebra. We also discuss related conjectures.

Consider the sum $a_1 + a_2 + a_3 + \cdots + a_n$. If we add brackets to this sum, as one usually does to evaluate a sum, one never adds the brackets in such a way that the symbols are added more than two at a time. The result is called an *arranged sum* or an *arrangement*. For example, $a_1 + a_2 + a_3 + a_4$ can be written as an arranged sum

$$(\alpha) \quad [(a_1 + a_2) + (a_3 + a_4)]$$

or

$$(\beta) \quad \{[(a_1 + a_2) + a_3] + a_4\}$$

We can define a *partial* sum to be the sum within any pair of brackets; e.g., in (β) the partial sums are

$$(a_1 + a_2), \quad (a_1 + a_2 + a_3), \quad (a_1 + a_2 + a_3 + a_4)$$

In (α) the partial sums are

$$(a_1 + a_2), \quad (a_3 + a_4), \quad (a_1 + a_2 + a_3 + a_4)$$

Conjecture C_{19} Given two arrangements of n symbols $a_1, \ldots, a_n$, one can choose nonzero values in Z_4 for the a_i in such a way that every partial sum of either arrangement is also nonzero.

For example, we could take $a_1 = 1$, $a_2 = 1$, $a_3 = 1$, and $a_4 = 2$ to satisfy the conjecture for the arrangements (α) and (β) above.

In order to relate this conjecture to the 4CC, we first introduce another seemingly number-theoretic conjecture which we shall use as an intermediary.

An *n-admissible set* is a set of ordered pairs of integers (p_i, q_i), $i = 1, \ldots, s$, such that

$$0 \leq p_i, p_i + 2 \leq q_i, \quad q_i \leq n \quad (i = 1, \ldots, s) \tag{5-8}$$

$$p_i < p_j < q_i < q_j \quad \text{is true for no } i \text{ and } j \tag{5-9}$$

An n-admissible set is *complete* if no new ordered pairs can be added without violating either (5-8) or (5-9). The geometric content of this notion is brought out by our next result.

Theorem 5-5 An n-admissible set (p_i, q_i), $1 \leq i \leq s$, corresponds to a set of inner diagonals in an $(n + 1) -$ gon P with vertices $0, \ldots, n$.

PROOF Let (p_i, q_i), $1 \leq i \leq s$, be an n-admissible set. By (5-9), each pair (p_i, q_i) corresponds to a nonconsecutive pair of vertices in P (except for the one exceptional case $(0, n)$). Condition (5-8) insures that the inner diagonals represented by the (p_i, q_i) never cross. Of course, we can replace "inner" by "outer" in the above theorem.

Corollary 5-2 A complete n-admissible set corresponds to a set of diagonals triangulating the interior (or exterior) of P.

Corollary 5-3 The cardinality of any complete n-admissible set is exactly $n - 2$.

Conjecture C_{20} If $(p_1, q_1), \ldots, (p_s, q_s)$ and $(p'_1, q'_1), \ldots, (p'_s, q'_s)$ are any two complete n-admissible sets, then it is always possible to find numbers $b_0, \ldots, b_n$ each equal to 1, 2, 3, or 4 such that

$$b_i \neq b_{i+1} \quad (i = 0, \ldots, n - 1) \tag{5-10}$$

$$b_{p_i} \neq b_{q_i}, \quad b_{p'_i} \neq b_{q'_i} \quad (i = 1, \ldots, n) \tag{5-11}$$

Theorems 5-6 Conjectures C_{20} and C_8 are equivalent.

PROOF By Corollary 5-2 Conjecture C_{20} is precisely the assertion that it is possible to assign colors $b_0, \ldots, b_n$ to the $n + 1$ vertices $0, \ldots, n$ of a polygon P whose interior and exterior are both triangulated by diagonals, so that no two consecutive vertices and no two vertices joined by a diagonal receive the same colors. But this is exactly Conjecture C_8.

Note that the constraint $b_0 \neq b_n$ is actually contained within (5-11), not (5-10). Now we can show that the earlier Conjecture C_{19} is also equivalent to the 4CC.

Lemma 5-2 Any arrangement of an n-fold sum gives rise to an n-admissible set, and vice versa.

PROOF Any related pair of parentheses in the arrangement encloses a sequence $a_r + \cdots + a_s$, and we associate to this pair of parentheses the ordered pair $(r - 1, s)$. It is easy to check that the set of all ordered pairs corresponding to the various related pairs of parentheses satisfies (5-8) and (5-9) and hence forms an n-admissible set. The converse is left as an exercise.

Note that the four-admissible set corresponding to (α) is $\{(0,2), (2,4), (0,4)\}$, while the four-admissible set corresponding to (β) is $\{(0,2), (0,3), (0,4)\}$.

Theorem 5-7 Conjectures C_{19} and C_{20} are equivalent.

PROOF Consider two arrangements of the n-fold sum $a_1 + \cdots + a_n$. Extend the analogous two n-admissible sets to make them complete—say $\{(p_i, q_i) \mid 1 \le i \le s\}$ and $\{(p'_i, q'_i) \mid \le i \le s\}$. If Conjecture C_{20} holds, then we can choose $b_0, \ldots, b_n$ so that (5-10) and (5-11) are satisfied. To show that Conjecture C_{19} is valid, we must find appropriate values for the a_i. Set $a_i \equiv b_i - b_{i-1} \pmod 4$. By (5-10) $a_i \not\equiv 0 \pmod 4$. Furthermore, any partial sum $a_u + \cdots + a_v$ for either arrangement corresponds to a related pair of parentheses and hence to an ordered pair $(u-1, v)$. By (5-11) $b_{u-1} \ne b_v$ or, equivalently, $b_v - b_{u-1} \ne 0$ (mod 4). But

$$a_u + \cdots + a_v \equiv (b_u - b_{u-1}) + (b_{u+1} - b_u) + \cdots + (b_v - b_{v-1})$$
$$= b_v - b_{u-1} \not\equiv 0 \pmod 4$$

Thus, none of the partial sums $a_u + \cdots + a_v$ (for either arrangement) is divisible by 4 so Conjecture C_{19} holds.

Conversely, if C_{19} holds, consider two complete n-admissible sets $\{(p_i, q_i) \mid 1 \le i \le s\}$ and $\{(p_i, q_i) \mid 1 \le i \le s\}$. By Lemma 5-2, there are two corresponding arrangements of the n-fold sum $a_1 + \cdots + a_n$, and so, by Conjecture C_{19}, we can find values for the a_i so that none of the partial sums for either arrangement are divisible by 4. Let $b_0 = 1$ and for $1 \le i \le n$, let $b_i \equiv 1 + a_1 + \cdots + a_i \pmod 4$ with $1 \le b_i \le 4$. Now $b_i - b_{i-1} \equiv a_i \not\equiv 0 \pmod 4$ so $b_i \ne b_{i-1}$ for $1 \le i \le n$. Thus, (5-10) is satisfied. On the other hand, if (u, v) belongs to one of the above n-admissible sets, the analogous partial sum $a_{u+1} + \cdots + a_v$ is not divisible by 4. Then $b_v - b_u \equiv a_{u+1} + \cdots + a_v \not\equiv 0$ (mod 4) so $b_u \ne b_v$. Hence, (5-11) holds and, therefore, Conjecture C_{20} holds.

Jan Mycielski has communicated to us the following conjecture which is equivalent to C_{20}. Let B_1 and B_2 be any two ways to insert parentheses in the sequence $x_1, x_2, \ldots, x_n$. Let us define a multiplication in the set $\{0, a, b, c\}$ by putting

$$x^2 = 0, \quad 0x = x0 = 0, \quad ab = c, \quad ac = b, \quad bc = a, \quad \text{and} \quad xy = yx$$

Conjecture C_{21} There exists a substitution of the letters $x_1, \ldots, x_n$ by the letters a, b, c such that $B_1 \ne 0$ and $B_2 \ne 0$.

Another numerical equivalent to the 4CC was given by Descartes and Descartes (1968). A *cartesian sequence* is a finite sequence $c(0), c(1), \ldots,$ of four colors such that

(a) $c(r) \ne c(r+1), r = 0, 1, 2, \ldots,$

i.e., the same color never appears in two consecutive positions, and

   (b)   the subsequence $c(2r)$, $r = 0, 1, 2, \ldots$, also enjoys property (a)

Conjecture C_{22} Given any integer n and an arbitrary increasing sequence of integers $0 \le i_0 < i_1 < \cdots < i_m \le n$, there exists a cartesian sequence $c(s)$, $s = 0, 1, 2, \ldots, n$, such that the subsequence $d(s) = c(i_s)$ is also cartesian.

Descartes and Descartes (1968) sketch a proof that C_{22} implies C_0 and state that the converse holds as well.

Still another numerical version of the 4CC was obtained by Bernhart (1972). For every $n \ge 2$, let us recursively define sets $X(n)$ and $Y(n)$ whose elements are integer sequences of length n, by setting $X(2) = (0, 0) = Y(2)$ ($(0, 0)$ is called *degenerate*) and requiring that each element $x \in X(n + 1)$ or $y \in Y(n + 1)$ be obtained from $x' \in X(n)$ or $y' \in Y(n)$ by inserting an element e between two consecutive elements of x' or y' and adding e to each of these elements, where $e = 1$ for x' and $e = 1$ or 2 for y'. In the case of y', the addition is to be performed modulo 3. If $y \in Y(n)$, then interchanging 1 and 2 produces a new $y' \in Y(n)$. We do not distinguish between y and y'. Thus,

$$X(3) = \{(1, 1, 1)\} = Y(3)$$
$$X(4) = \{(1, 2, 1, 2), (2, 1, 2, 1)\}$$
$$Y(4) = \{(1, 2, 1, 2)(1, 0, 2, 0), (0, 1, 0, 2)\}$$

If $w \in W(n + 1)$ is obtained from $w' \in W(n)$ by inserting e as above between the ith and $i + 1$st entries, we write $w = w'_i$ ($W = X$ or Y). We recursively define a binary relation x/y for $x \in X(n)$ and $y \in Y(n)$: $(0, 0)/(0, 0)$ and if $x \in X(n + 1)$, $y \in Y(n + 1)$ for $n \ge 2$, then x/y if and only if there exist $x' \in X(n)$, $y' \in Y(n)$, and i ($1 \le i \le n - 1$) such that $x = x'_i$, $y = y'_i$, and x'/y'.

Now let us state Bernahrt's conjecture.

Conjecture C_{23} For all $n \ge 2$, if x and $\bar{x} \in X(n)$, then there exists $y \in Y(n)$ such that x/y and $\bar{x}/y$.

It is worth remarking that this conjecture and its connection with the 4CC have a geometric origin. If M is a cubic map whose dual has no separating triangles, then Whitney's or Tutte's theorems (see Appendix) guarantee, the existence of a Hamiltonian circuit H in $D(M)$. Superimposing H on M, we find that H cuts a certain number of edges. Each edge $e = [v, w]$ cut by H is split in the middle to produce two new edges $[v, v_e]$ and $[w, w_e]$. It is not difficult to see that if every edge $e = [v, w]$ cut by H is replaced by $[v, v_e]$, $[w, w_e]$ and all other edges and vertices are left alone, then two trees are produced. Vertices of degree 1 in these trees are just v_e or w_e for some e and so may be paired. Elements of X are produced by examining these trees; elements of Y arise from applying Heawood vertex characters.

5-6 FLOW RATIOS

For any graph G, let $\mathscr{D}(G) = \{D \mid D$ is a directed graph and $U(D) = G\}$ and let $\mathscr{A}(G) = \{D \in \mathscr{D}(G) \mid D$ is acyclic$\}$. Let $A(D)$ be the set of arcs of D.

Lemma 5-3 For every G, $\mathscr{A}(G)$ is nonempty.

PROOF. List the vertices of G in any order—say, $v_1, \ldots, v_n$. Let $(v_i, v_j) \in A(D)$ if and only if $[v_i, v_j] \in E(G)$ and $i < j$. Clearly, $D \in \mathscr{D}(G)$.

If D is an acyclic graph, we define, as in Berge (1973, p. 364), $t(D)$ to be the length (number of arcs) of the longest path in D. The next result is due to Gallai (1968).

Theorem 5-8 If $D \in \mathscr{A}(G)$, then $\chi(G) \leq 1 + t(D)$.

PROOF If $v \in V(D)$, set $t(v) = $ length of longest path in D beginning at v. Define a coloring c of G by setting $c(v) = 1 + t(v)$. Since $t(D) = \max\{t(v) \mid v \in V(D)\}$, c is a $t(D)$-coloring of G.

Corollary 5-4 For any graph G, $\chi(G) = \inf\{1 + t(D) \mid D \in \mathscr{A}(G)\}$.

PROOF By Theorem 5-8, it suffices to find $D \in \mathscr{A}(G)$ such that $\chi(G) \geq 1 + t(D)$. But this is easily done. If c is any $\chi(G)$-coloring of G using colors $1, 2, \ldots, \chi(G)$, then direct an edge $e = [v, w]$ from v to w if $c(v) > c(w)$ and from w to v if $c(w) > c(v)$. The resulting directed graph $D \in \mathscr{A}(G)$ can clearly have no paths of length greater than $\chi(G) - 1$.

Of course, we have now proved the equivalence of the following conjecture with the 4CC.

Conjecture C_{24} If G is any planar graph, then there exists $D \in \mathscr{A}(G)$ with $t(D) \leq 3$.

Corollary 5-4 and its proof have an interesting consequence which strengthens the easy observation that $\chi(G)$-coloring of G is necessarily a "complete" coloring (see Sec. 6-2).

Corollary 5-5 Let G be any graph with $\chi(G) = k$ and let c be any k-coloring of G. If $V_1, \ldots, V_k$ are the color classes of G with respect to c (so $V_i = c^{-1}(i)$), then, for any permutation $\pi_1, \ldots, \pi_k$ of $1, \ldots, k$, there is a path $(w_1, \ldots, w_k)$ in G with $w_i \in V_{\pi_i}$.

PROOF We give the proof when $\pi_j = j$, $1 \leq j \leq k$; the modification for an arbitrary permutation is trivial and is left as an exercise. If D is the directed graph constructed in the proof of Corollary 5-4 we simply note that, by

Theorem 5-8, $1 + t(D) = \chi(G)$ and so D contains a path of length $\chi(G) - 1$. But any such path must consist of vertices $w_1, \ldots, w_k$ colored consecutively $1, \ldots, k$, and hence $w_i \in V_i$, $1 \le i \le k$.

If D is a directed graph, D is *transitive* if (u, v) and (v, w) in $A(D)$ (the set of arcs of D) imply $(u, w) \in A(D)$ for all u, v, $w \in V(D)$ (the set of vertices of D). Any acyclic digraph D has a *transitive closure* $T(D)$ defined as follows. The vertices of $T(D)$ are just those of D; two vertices v, w are joined by an arc (v, w) in $T(D)$ if and only if there is a directed path from v to w in D.

The proof of the following lemma is easy.

Lemma 5-4 For any acyclic digraph D, $t(D) = t(T(D))$.

As an immediate consequence of this lemma and of Corollary 5-4 we have the following corollary.

Corollary 5-6 For any graph G,

$$\chi(G) = \inf\{1 + t(T(D)) \mid D \in \mathscr{A}(G)\}$$

Let $U(D)$ be the undirected graph associated with D.

We shall now derive a theorem of Minty (1962). If D is an acyclic directed graph and C is a circuit in $U(D)$, the *flow ratio* $f(C)$ is the ratio m/n of the number m of arcs of C in one direction to the number n of arcs in the other direction, where $m \ge n$. Since D is acyclic, $n \ne 0$ and so $f(C)$ is finite. Let $f(D) = \max\{f(C) \mid C$ a circuit of $U(D)\}$.

Lemma 5-5 Let D be an acyclic graph. Then $f(D) \le t(D)$.

PROOF Choose a circuit C in $U(D)$ with $f(C) = f(D)$. One can show that $f(C) \le t(C)$ and plainly $t(C) \le t(D)$. Thus, $f(D) \le t(D)$.

Theorem 5-9 $\chi(G) = \min\{1 + f(D) \mid D \in \mathscr{A}(G)\}$.

PROOF If $D \in \mathscr{A}(G)$ is the directed graph obtained from a $\chi(G)$-coloring of G, then by Corollary 5-4 and Lemma 5-5

$$\chi(G) = 1 + t(D) \ge 1 + f(D) \ge \min\{1 + f(D) \mid D \in \mathscr{A}(G)\}.$$

To finish the proof, we must show $\chi(G) \le k$ provided that $f(C) \le k - 1$ for every circuit C in some $D \in \mathscr{A}(G)$.

We may suppose that G is connected. Select a starting vertex v_0 and color it with the color 1 and consider some other vertex v_p. We shall define an auxiliary integer-valued function $g(v_p)$ whose values when reduced mod k yield the desired coloring. As a preliminary operational step for defining $g(v_p)$, we define the gain of a chain of arcs, when traversed from v_0 to v_p, to be the number of arcs traversed in their inherent direction minus $k - 1$ times the number of arcs traversed in the opposite sense.

To show that such a chain with maximum gain exists, note first that if a chain is not simple and hence contains a circuit, the circuit cannot contribute to the gain because of the flow ratio condition, and hence it may be deleted from the chain to yield a chain with equal or greater gain. Again, if the resulting chain is not simple, a circuit is deleted, and so on until a simple chain is obtained. Since the number of simple chains is finite, there is one with maximum gain. If with each vertex v_p is associated the maximum gain $g(v_p)$ for a chain oriented from v_0, then we have for two vertices v_p and v_q joined by an arc

$$0 \le \left| g(v_p) - g(v_q') \right| < k$$

For if $g(v_p) > g(v_q)$, then $g(v_q) \ge g(v_p) - (k - 1)$ so $\left| g(v_p) - g(v_q) \right| \le k - 1$. Thus, $g(v_p)$ when reduced mod k yields a k-coloring of the vertices, since the values of g for adjacent vertices differ by less than k and hence cannot produce the same integer mod k.

It is now obvious that the following conjecture is equivalent to C_1.

Conjecture C_{25} If G is planar, then there exists $D \in \mathscr{A}(G)$ such that $f(D) \le 3$.

Note that, in general, $f(D) \ne t(D)$. The reader can easily orient the edges of a circuit with G vertices to produce a directed graph D for which $t(D) = 4$ but $f(D) = 2$.

Corollary 5-7 If $D \in \mathscr{A}(G)$ is obtained from a $\chi(G)$-coloring of G, then $t(D) = f(D)$.

PROOF $1 + t(D) = \chi(G) \le 1 + f(D) \le 1 + t(D)$ by Corollary 5-4, Lemma 5-5, Theorem 5-9.

The converse is false since if the edges of the four-vertex circuit C_4 are oriented so that, for the resulting D, $t(D) = 3$, then $f(D) = 3$ also. But, of course, $\chi(C_4) = 2$.

CHROMATIC NUMBERS AND CHROMATIC POLYNOMIALS

6-1 INTRODUCTION

In this chapter, we have collected a number of important approaches to the general problem of coloring which began as attempts to solve the 4CC. Section 6-2 covers Brooks' theorem and Wilf's theorem, as well as other generalities on the smallest number of colors needed for arbitrary graphs. Then, in the next section, we move from chromatic number to chromatic index and the theorems of Vizing and Shannon. After introducing the concept of chromatic polynomials and developing their basic properties in Sec. 6-4, the next two sections concentrate on the work of Tutte (on the Golden Root) and Whitney. The intriguing ideas of Berman and Tutte and of Read, dealing with the behavior of chromatic polynomials, are also presented. In Sec. 6-7, we give the Birkhoff–Lewis conjecture (a strengthened form of the 4CC) and in Sec. 6-8 the Beraha numbers are discussed, along with the theory of unconstrained and constrained chromials by Hall.

6-2 BROOKS' THEOREM AND OTHER BOUNDS

An *independent set* W in a graph G is a subset W of $V(G)$ such that no two vertices in W are adjacent. A *partition* of $V(G)$ is any collection $W_1, \ldots, W_k$ of disjoint nonempty subsets W_i of $V(G)$ such that $V(G) = W_1 \cup \cdots \cup W_k$. The *chromatic number* $\chi(G)$ of G is the minimum number of elements in any partition of $V(G)$ into independent sets. The reader can easily verify that this is equivalent to the definition of chromatic numbers as the smallest integer k for which G can be k-colored, as given in Chap. 2. For completeness, let us state the following version of Conjecture C_1.

Conjecture C_{26} For any planar graph, $\chi(G) \leq 4$.

Let $\alpha(G)$ be the largest number of vertices in any independent set. The following lemma is an exercise for the reader.

Lemma 6-1 For any graph G with n vertices, $n/\alpha(G) \leq \chi(G) \leq n - \alpha(G) + 1$.

Note that the bounds of Lemma 6-1 are not, in general, attainable. For example, if G is the graph of Fig. 6-1, then $n = 6$ and $\alpha(G) = 3$. Thus, $\frac{6}{3} = 2 < \chi(G) < 6 - 3 + 1 = 4$; since $\chi(G) = 3$, both inequalities are strict.

Let $\omega(G)$ denote the maximum order of any complete subgraph of G. Clearly, we have the following.

Lemma 6-2 For all G, $\chi(G) \geq \omega(G)$.

If G is simple, let $\overline{G}$ denote the complement of G; $V(\overline{G}) = V(G)$ and $[v, w] \in E(\overline{G})$ if and only if $[v, w] \notin E(G)$. Plainly, $\alpha(G) = \omega(\overline{G})$. This suggests defining $\bar{\chi}(G) = \chi(\overline{G})$; thus, $\bar{\chi}(G)$ is just the minimum number of elements in a cover of G by subsets W_i, each of which induces a complete subgraph $G(W_i)$ of G. The notation $\theta(G)$ may also be used instead of $\bar{\chi}(G)$.

Lemma 6-3 For all G, $\bar{\chi}(G) \geq \alpha(G)$.

PROOF Assume without loss of generality that G is simple. Now apply Lemma 6-2 to $\overline{G}$.

A graph G is called χ-*perfect* if $\chi(G') = \omega(G')$ for all full subgraphs G' of G; G is α-*perfect* if $\bar{\chi}(G') = \alpha(G')$ for all full subgraphs G' of G. The following

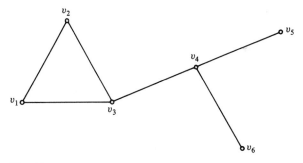

Figure 6-1

theorem settled in the affirmative a well-known conjecture of Berge regarding these two notions.

Theorem 6-1 (Lovasz, 1972) G is χ-perfect if and only if G is α-perfect.

Lovasz's proof uses the theory of "hypergraphs" (see Berge, 1973).

One cannot hope, in general, to obtain any nice relation between $\chi(G)$ and $\omega(G)$ since there are graphs with arbitrarily high chromatic number but no triangles. The construction we present is due to Descartes (1954). In fact, it yields, for every $k > 2$, a graph with girth at least 6 and chromatic number at least k.

Let $G_3 = C_7$, the circuit with seven vertices. For $i \geq 3$, we recursively define G_{i+1} in terms of G_i. Suppose G_i has n_i vertices. Take

$$w_i = \binom{i(n_i - 1) + 1}{n_i}$$

disjoint copies of G_i, together with $i(n_i - 1) + 1$ new "central" vertices. Each of the w_i distinct copies of G_i is thus "indexed" by an n_i-element subset of the central vertices. Join each of the w_i copies of G_i to its corresponding n_i-element subset of the central vertices by n_i disjoint edges. The resulting graph is G_{i+1}. It is fairly easy to check that the girth of each G_i is at least 6 and inductively that $\chi(G_k) \geq k$.

Here is a simple example of a graph G with no triangles (Fig. 6-2) but with $\chi(G) = 4$.

According to Lemma 6-1, if the 4CC is true, then, for any planar graph G with n vertices, $\alpha(G)/n \geq \frac{1}{4}$. This has been called the *Erdös–Vizing conjecture*. Of course, it follows from the five-color theorem that $\alpha(G)/n \geq \frac{1}{5}$ and Albertson (1974) proved that $\alpha(G)/n > \frac{2}{9}$.

Let us return to the invariants $\chi(G)$ and $\bar{\chi}(G)$. We have the following results of Nordhaus and Gaddum (1956).

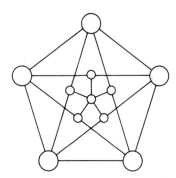

Figure 6-2

Theorem 6-2 For any graph G with n vertices:

(a) $2\sqrt{n} \leq \chi(G) + \bar{\chi}(G) \leq n + 1$

(b) $\quad n \leq \chi(G)\bar{\chi}(G) \leq \left(\dfrac{n+1}{2}\right)^2$

PROOF Since $\bar{\chi} \leq \alpha$, the first inequality of (b) follows from the first inequality of Lemma 6-1. We can get the first inequality of (a) by a formal manipulation. Let $t = \frac{1}{2}(\chi(G) + \bar{\chi}(G))$ be the arithmetic mean of $\chi(G)$ and $\bar{\chi}(G)$. Then $\chi(G) = t + a$ and $\bar{\chi}(G) = t - a$ for some real number a. Hence, $n \leq \chi(G)\bar{\chi}(G) = t^2 - a^2 \leq t^2$, so $t \geq \sqrt{n}$. Thus, $\chi(G) + \bar{\chi}(G) = 2t \geq 2\sqrt{n}$.

Now let us establish the second inequality of (a). We use induction on the number n of vertices. For $n = 1$, the result is trivial. Suppose it holds for all G with at most $n - 1$ vertices and let H be an arbitrary graph with n vertices ($n \geq 2$). Pick $v \in V(H)$ and put $G = H - v$. Clearly,

$$\chi(H) \leq \chi(G) + 1$$

It is easy to check that $\bar{G} = \bar{H} - v$. Hence, we also have

$$\bar{\chi}(H) \leq \bar{\chi}(G) + 1$$

Moreover, by the inductive hypothesis, if either of the inequalities above is a strict inequality, then $\chi(H) + \bar{\chi}(H) \leq n + 1$. Suppose then that $\chi(H) = \chi(G) + 1$ and $\bar{\chi}(H) = \bar{\chi}(G) + 1$. It follows that the degree d of v in H is at least $\chi(G)$ since, if $d < \chi(G)$, any $\chi(G)$-coloring of G extends to H. Similarly, $\bar{d} \geq \bar{\chi}(G)$, where $\bar{d} = n - d - 1$ is the degree of v in $\bar{H}$. Thus, $\chi(H) + \bar{\chi}(H) = \chi(G) + 1 + \bar{\chi}(G) + 1 \leq d + 1 + n - d - 1 + 1 = n + 1$, as required.

The second inequality in (b) is an immediate consequence of this result together with the fact that the product of two numbers with bounded sum is maximized when the numbers are equal.

Let us close this section with a short and elegant proof by Lovász (1974) of Brooks' theorem.

Theorem 6-3 If $\Delta(G) \leq k$ ($k \geq 3$) and G contains no K_{k+1}, then G is k-colorable.

PROOF First note that the theorem, as given above, is equivalent to the statement of Theorem 2-9. Suppose G has n vertices. Let us assume that G contains two vertices a and b such that $d(a, b) = 2$ and $G - a - b$ is connected. (We shall justify the assumption in a moment.)

Let v be a vertex adjacent to both a and b. List the $n - 2$ vertices of $G - a - b$ beginning with v, $v = v_1, v_2, \ldots, v_{n-2}$ so that each vertex v_i ($i \geq 2$) is adjacent to some v_j with $j < i$. (The verification that this is always possible is an exercise.)

We color G as follows. Color vertices a and b with color 1; this is legitimate since $d(a, b) = 2$ implies a and b are not adjacent. Now color the vertices $v_1, \ldots, v_{n-2}$ in *reverse order*; i.e., start by coloring v_{n-2}, then v_{n-3}, v_{n-4}, etc. At each stage, the vertex v_i to be colored is adjacent to at least one v_j with $j < i$ (for $i \geq 2$) and thus because $\deg(v_i) \leq k$, v_i has at most $k-1$ already-colored neighbors v_r $(r > i)$. The vertex $v_1 = v$ is adjacent to a and b, both colored 1, and the other $\leq k-2$ neighbors of v can use at most $k-2$ additional colors. This still leaves at least one free color for v.

Let us justify our original assumption. Clearly, we may assume G is two-connected. (Why?) Since G is not a complete graph, we can choose a vertex x not adjacent to all the other vertices. If $G - x$ is two-connected, let $a = x$ and take b any vertex at distance 2 from x. If $G - x$ is not two-connected, then the block-cutpoint tree $bc(G - x)$ of $G - x$ is nontrivial. Choose two endpoints of $bc(G - x)$ and let A and B be the corresponding blocks of $G - x$. Since G is two-connected, x must be adjacent to points $a \in A$ and $b \in B$, which cannot be adjacent themselves. This completes the justification.

Since the line graph of a cubic graph is regular of degree 4, Brooks' theorem implies that any cubic graph can be edge-colored with four colors and this, in turn, implies the earlier result of Theorem 4-4.

Wilf (1967) has found a very different kind of bound on $\chi(G)$, similar in form but not substance to Brooks' theorem. Since the proof would take us very far afield, we merely state the result.

Theorem 6-4 If G is a connected simple graph, then

$$\chi(G) \leq 1 + \varepsilon(G)$$

where $\varepsilon(G)$ denotes the maximum (real) eigenvalue of the adjacency matrix of G. Moreover, equality holds if and only if G is a complete graph or an odd circuit.

Call a coloring c of G *essential* (or *complete*) if for every pair of distinct colors c_i, c_j, there is a pair v_i, v_j of adjacent vertices with $c(v_i) = c_i$ and $c(v_j) = c_j$. Now, while a graph G has r-colorings for all r sufficiently large, they will not all be essential colorings. For example, C_4 has no essential r-colorings for $r > 2$. In general, G has no essential r-colorings for $r > |V(G)|$. This suggests a definition: the *achromatic number* $\psi(G)$ of G is the largest r for which G has an essential r-coloring. Since a nonessential r-coloring induces an $(r-1)$-coloring (by merging two nonadjacent color classes), a $\chi(G)$-coloring of G is certainly essential. Hence, we have

Lemma 6-4 $\psi(G) \geq \chi(G)$.

It follows immediately that $\psi(K_n) = \chi(K_n) = n$. The chromatic and achromatic numbers are also the same for complete bipartite graphs.

Lemma 6-5 $\psi(K_{r,s}) = 2$.

PROOF Let c be any coloring of $K_{r,s}$ with colors $c_1, \ldots, c_t$ and let $V = X \cup Y$ be a partitioning of $V = V(K_{r,s})$ into two independent sets. Since $K_{r,s}$ is complete, any two vertices $x \in X$ and $y \in Y$ are adjacent. Hence, no color c_i appears in both X and Y. If c is a complete coloring, every two color classes contain a pair of adjacent vertices. Thus, no two color classes can be entirely contained within X or within Y. The result now follows.

The two examples notwithstanding, achromatic and chromatic numbers are quite different. In particular, removing edges from a graph G cannot increase $\chi(G)$ but it can increase $\psi(G)$. For example, C_6 is a spanning subgraph of $K_{3,3}$, and the following lemma shows $\psi(C_6) = 3$ while $\psi(K_{3,3}) = 2 = \chi(K_{3,3}) = \chi(C_6)$.

Lemma 6-6 $\psi(C_6) = 3$.

PROOF Figure 6-3 shows that $\psi(C_6) \geq 3$. If C_6 had a complete four-coloring, one of the color classes would consist of a single vertex v. But v has degree 2 and hence cannot be adjacent to the three other color classes.

6-3 CHROMATIC INDEX

If G is a graph, we define the *chromatic index* $\chi'(G)$ by the formula $\chi'(G) = \chi(LG)$, where LG is the line graph of G. Thus, $\chi'(G)$ is the minimum number of colors needed to color the edges of G in such a way that no two adjacent edges have the same color. Since $\omega(LG) = \Delta(G)$, we have the following lemma.

Lemma 6-7 $\chi'(G) \geq \Delta(G)$.

An important result of Vizing (1964) (see also Ore, 1967, p. 245) says that $1 + \Delta(G)$ colors will always suffice to color the edges of G if G has no multiple edges, and that, more generally, we have the following theorem whose proof will not be given.

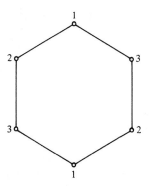

Figure 6-3

Theorem 6-5 $\chi'(G) \leq s + \Delta(G)$, where s is the maximum number of edges joining any pair of vertices.

There is a different upper bound due to Shannon (1949) which can be derived from Vizing's theorem. Our proof essentially follows that of Berge (1973). In a graph G, $m(v, w)$ denotes the number of edges joining the pair v, w of vertices; clearly, $s = \max\limits_{v,w} m(v, w)$.

Theorem 6-6 For any loopless graph G

$$\chi'(G) \leq \frac{3\Delta(G)}{2}$$

PROOF Suppose G is a counter-example to the theorem. If $s \leq \Delta(G)/2$, then by Vizing's theorem, $\chi'(G) \leq \Delta(G) + (\Delta(G)/2) \leq 3\Delta(G)/2$. Hence, $m(v, w) \geq (\Delta(G)/2) + 1$ for some pair v, w of vertices in G. We may assume without loss of generality that the removal of any edge from G produces a graph G' with $\chi'(G') \leq 3\Delta(G)/2$.

Now choose an edge e in G joining the vertices v and w given above and consider the corresponding vertex $\bar{e}$ in $L(G)$. Let k be the degree of $\bar{e}$ in $L(G)$; this is just the number of edges in G adjacent to e. Hence,

$$k \leq (\deg(v) - 1) + (\deg(w) - 1) - (m(v, w) - 1)$$

(The last summand $m(v, w) - 1$ represents the edges other than e which join v, and w that are counted twice, both at v and at w.) Now $\deg(v) \leq \Delta(G)$, $\deg(w) \leq \Delta(G)$, and $m(v, w) - 1 \geq \Delta(G)/2$, so

$$k \leq 2\Delta(G) - 2 - \frac{\Delta(G)}{2} \leq \frac{3\Delta(G)}{2} - 1$$

Hence, any $(3\Delta(G)/2)$-edge-coloring of $G - e$ extends to G, violating the assumption that G was a counter-example to the theorem.

The key property of G in the above proof can be described as being "edge-critical" with respect to edge-coloring. See the Appendix for the general notion of critical graphs.

Corollary 6-1 Let G be a graph. Then $\chi'(G) \leq \Delta(G) + \min(s, \Delta(G)/2)$. $\chi'(G) \leq 4$ if G is a cubic graph can now be derived again.

Conjecture C_{27} If G is a bridgeless planar cubic graph, then $\chi'(G) = 3$. This is simply a restatement of Conjecture C_4.

REMARK Let $\langle x \rangle$ be the *fractional part* of x, $\langle x \rangle = 0$ if x is an integer, and $\langle x \rangle = 1$ if x is not an integer. Recall that $\{x\} = \inf\{n \mid n \geq x \text{ and } n \text{ an integer}\}$

and $[x] = \sup\{n \mid n \le x$ and n an integer$\}$. Note that $\{x\} = -[-x]$ and that $\langle x \rangle = \{x\} - [x]$.

Let G be any connected simple graph with n vertices and m edges. Let $t = (n^2 - 2m)/n$. Ershov and Kozhukhin (1962) have proved that $l \le \chi(G) \le u$, where

$$l = \left\{ \frac{n}{[t]} \left(1 - \frac{\langle t \rangle}{1 + [t]} \right) \right\} \quad \text{and} \quad u = \frac{3 + \sqrt{9 + 8(m - n)}}{2}$$

For example, if t is an integer, $l = \{n/t\}$. It is easy to check that:
(a) If $G = K_n$, $l = n = u$.
(b) If G is a tree, $l = 2 = u$.

6-4 THE NUMBER OF WAYS TO COLOR A GRAPH; CHROMATIC POLYNOMIALS

In the remainder of this chapter, we shall present the basic facts about chromatic polynomials, including the seminal work of Whitney and Tutte. We have drawn upon several excellent expositions of the theory given by Tutte (1969b, 1970a, 1970b) and by Read (1968).

Let G be a graph and let t be a nonnegative integer. We denote by $P(G, t)$ the number of colorings of G using t or fewer colors. If c is such a coloring of G from t colors, then c can be regarded as a function from $V(G)$ to a set with t elements—say $\{1, \ldots, t\}$—so that $c(v) \ne c(w)$ whenever v is adjacent to w. Two colorings of G are regarded as different even if they differ only by a (nonidentity) permutation of the colors. It will be convenient to admit the *empty graph* Ω, which is the graph with no vertices.

Now $P(G, 0) = 0$ unless $G = \Omega$, and $P(\Omega, t) = 1$ for any t. It is easy to see that $P(K_1, t) = t$, $P(K_2, t) = t(t - 1)$ and $P(\overline{K}_2, t) = t^2$. If G contains a loop, then $P(G, t) = 0$. Obviously, the deletion of multiple edges does not affect $P(G, t)$.

The following lemmas are obvious.

Lemma 6-8 $P(\overline{K}_n, t) = t^n$.

Lemma 6-9 $P(K_n, t) = t_{(n)} = t(t - 1)\cdots(t - n + 1)$.

If G has two connected components H and K, clearly $P(G, t) = P(H, t)P(K, t)$. Our next result generalizes this.

Theorem 6-7 Let G be the union of two subgraphs H and K whose intersection L is a complete graph K_n. (Note that if $n = 0$, $K_n = \Omega$). Then

$$P(G, t) = \frac{P(H, t)P(K, t)}{P(L, t)}$$

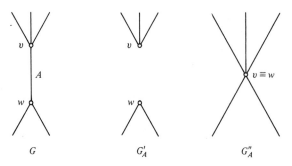

G G'_A G''_A

Figure 6-4

PROOF Let c_1 be any coloring of H and let c'_1 be the induced coloring of L. There are precisely $P(K,t)/P(L,t)$ different colorings c_2 of K for which $c'_1 = c'_2$, where c'_2 is the coloring of L induced by c_2. But a coloring of G is precisely a pair (c_1, c_2) of colorings of H and K, respectively, such that $c'_1 = c'_2$.

For example, if P_3 is the path with three vertices, then

$$P(P_3, t) = \frac{P(K_2, t)P(K_2, t)}{P(K_1, t)} = \frac{t^2(t-1)^2}{t} = t(t-1)^2$$

Using the same argument, the reader can show that $P(P_4, t) = t(t-1)^3$.

Suppose A is any edge of G. Denote by G'_A the graph obtained from G by deleting A and by G''_A the graph obtained by removing A and then identifying its endpoints (see Fig. 6-4). Note that if A is a loop, $G'_A = G''_A$.

Theorem 6-8 $P(G, t) = P(G'_A, t) - P(G''_A, t)$.

PROOF It is easier to show that $P(G'_A, t) = P(G, t) + P(G''_A, t)$. Let $A = [v, w]$. First suppose $v \neq w$. Then any coloring c of G'_A satisfies either (1) $c(v) \neq c(w)$ or (2) $c(v) = c(w)$. In case (1), c induces a coloring of G, and in case (2), a coloring of G''_A. On the other hand, if $v = w$, then A is a loop. Hence, $P(G, t) = 0 = P(G'_A, t) - P(G''_A, t)$.

We can use this theorem to compute $P(G, t)$ for a graph G with m edges in terms of its $(m-1)$-edge subgraphs. For example, suppose $G = C_4$, the circuit of length 4. Then we compute $P(G, t)$ as in Fig. 6-5.

Figure 6-5

The notation here (first given by Zykov, 1949) is intended to suggest pictorially the equation of Theorem 6-8. Thus, we may write

$$P(C_4, t) = P(P_4, t) - P(K_3, t)$$
$$= t(t-1)^3 - t(t-1)(t-2)$$
$$= t(t-1)[(t-1)^2 - (t-2)]$$
$$= t^4 - 4t^3 + 6t^2 - 3t$$

The following lemmas are easy computations; their proofs are left as exercises.

Lemma 6-10 $P(P_{n+1}, t) = t(t-1)^n$

Lemma 6-11 $P(C_n, t) = (t-1)^n + (-1)^n(t-1)$

In every instance, we have seen that $P(G, t)$ is a *polynomial function* of t. This property is, in fact, quite general and will be of extreme importance to us. We shall use an inductive argument due to Tutte to show that $P(G, t)$ is a polynomial in t and to establish some of its properties. Let k be the number of components.

Theorem 6-9 Let G be a graph without loops. Then $P(G, t)$ is a monic polynomial in t with integer coefficients and of degree $n = |V(G)|$. The coefficients of the terms of degree $\geq k$ are all nonzero and alternate in sign.

PROOF We induct on $m = |E(G)|$. If $m = 0$, the theorem is a trivial consequence of Lemma 6-8. Suppose $m > 0$ and that the theorem holds for $m - 1$. Consider G'_A and G''_A for some edge A of G. Both G'_A and G''_A have fewer than m edges and G'_A has no loops. If G''_A has a loop, then $P(G, t) = P(G'_A, t)$ and so we are done by induction since G'_A has order $n = $ order of G. On the other hand, if G''_A has no loops, then setting $P(G'_A, t) = t^n - \alpha'_1 t^{n-1} + \alpha'_2 t^{n-2} \cdots$ and $P(G''_A, t) = t^{n-1} - \alpha''_1 t^{n-2} + \cdots$, where the α'_i and α''_j are all positive, we have, by Theorem 6-8,

$$P(G, t) = t^n - \alpha'_1 t^{n-1} + \alpha'_2 t^{n-2} \cdots - t^{n-1} + \alpha''_1 t^{n-2} \cdots$$
$$= t^n - (\alpha'_1 + 1)t^{n-1} + (\alpha'_2 + \alpha''_1)t^{n-2} \cdots$$

from which the theorem follows immediately.

Thus, for any loopless graph $G \neq \Omega$ of order n, we can write

$$P(G, t) = \sum_{i=0}^{n} (-1)^i \alpha_i t^{n-i} \ (*)$$

where $\alpha_i > 0$ for $0 \leq i \leq n - k$ and $\alpha_0 = 1$. We sometimes refer to these coefficients as $\alpha_i(G)$ to make clear their dependence on G. Obviously, it is important to try to compute these coefficients. The following easy results are of this type.

Lemma 6-12 If $G \neq \Omega$ has order n, then $\alpha_n(G) = 0$.

PROOF $P(G,0) = \alpha_n(G)$. But $P(G,0) = 0$ unless $G = \Omega$.

Lemma 6-13 If G is a simple graph of order n with m edges, then $\alpha_1(G) = m$.

PROOF Exercise.

Let us now derive some results about $\alpha_i(G)$ for a connected graph G by using spanning trees. For the moment, we assume $P(T,t) = t(t-1)^{n-1}$ for a tree T of order n. (This result is proved in Lemma 6-14.)

Theorem 6-10 If G is connected of order n, then $P(G,t) \leq t(t-1)^{n-1}$

PROOF Let T be any spanning tree of G. Since every coloring of G induces a coloring of T, $P(G,t) \leq P(T,t) = t(t-1)^{n-1}$.

In fact, this inequality is an equality if and only if G is a tree.

Theorem 6-11 A simple graph G of order n is a tree if and only if $P(G,t) = t(t-1)^{n-1}$

PROOF We have already seen that $P(G,t) = t(t-1)^{n-1}$ if G is a tree. Conversely, suppose $P(G,t) = t(t-1)^{n-1}$. Then G is connected (else $P(G,t)$ is divisible by a higher power of t) and has $n-1$ edges by Lemma 6-13. Hence, G is a tree.

Let $H = J + A$ be obtained from a graph J of order n by adding an edge A. Then $H'_A = J$. By the proof of Theorem 6-9, $\alpha_i(H) = \alpha_i(H'_A) + \alpha_{i-1}(H''_A)$ so $\alpha_i(H) \geq \alpha_i(J)$ for $0 \leq i \leq n-1$.

Thus, we have the following theorem.

Theorem 6-12 Let H be a spanning subgraph of G. Then $\alpha_i(G) \geq \alpha_i(H)$ for $0 \leq i \leq n-1$.

By using the spanning tree method, we can now find a lower bound for the $\alpha_i(G)$.

Theorem 6-13 Let G be a connected graph of order n. Then $\alpha_{n-r}(G) \geq \binom{n-1}{r-1}$, $1 \leq r \leq n$, where $\binom{a}{b}$ denotes the binomial coefficient.

PROOF Let T be a spanning tree of G. By Theorem 6-12, $\alpha_i(G) \geq \alpha_i(T)$.

Using the binomial theorem, we get

$$P(T, t) = t(t - 1)^{n-1} = \sum_{r=1}^{n} (-1)^{n-r-1} \binom{n-1}{r-1} t^r$$

so

$$\alpha_{n-r}(T) = \binom{n-1}{r-1}$$

which suffices.

6-5 CHROMATIC POLYNOMIALS AND THE GOLDEN RATIO

By Theorem 6-9, we can define for any graph G a polynomial $P(G, \lambda)$, where λ is an indeterminate. $P(G, \lambda)$ is called the *chromatic polynomial* of G. Of course, $P(G, \lambda) \equiv 0$ if G has a loop and $P(\Omega, \lambda) \equiv 1$. The coefficients of $P(G, \lambda)$ are otherwise just $(-1)^i \alpha_i(G)$. A polynomial $P(\lambda)$ is chromatic (or a *chromial*) if $P(\lambda) = P(G, \lambda)$ for some graph G. Because of results such as Theorem 6-9, not every polynomial is a chromial. The problem of determining which polynomials are chromial seems to be very difficult. It is also not known when G and H have different chromatic polynomials except in very trivial cases. (Recently, Tutte (1973a) and Lee (1974) have systematically constructed a variety of such examples which are maximal planar.) We do have a characterization (Theorem 6-11) of trees in terms of their chromials, but to complete its proof we must still show the following.

Lemma 6-14 Let T be a tree with n vertices. Then $P(T, \lambda) = \lambda(\lambda - 1)^{n-1}$.

PROOF The lemma is clear for $n = 1$. Suppose it holds for $n - 1$ and let T be any tree with $n \geq 2$ vertices. Choose a vertex v of degree 1. Then $T - v$ is a tree with $n - 1$ vertices. Hence, by induction, $P(T - v, \lambda) = \lambda(\lambda - 1)^{n-2}$. Let $e = [v, w]$ be the unique edge of T incident with v. Then e determines a subgraph of T isomorphic with K_2 and so, by Theorem 6-7,

$$P(T) = \frac{\lambda(\lambda - 1)^{n-2} \lambda(\lambda - 1)}{\lambda} = \lambda(\lambda - 1)^{n-1}$$

By regarding $P(G, \lambda)$ as a polynomial, we can now ask about the behavior of $P(G, \lambda)$ on the real line (or even in the complex plane) and about the roots of $P(G, \lambda)$ (see Berman and Tutte, 1969). This suggests the following formulation of the 4CC, which is evidently equivalent to Conjecture C_1.

Conjecture C_{28} If G is planar, $\lambda = 4$ is not a root of $P(G, \lambda)$.

Let us explicitly remark here that we can, of course, define $P(M, \lambda)$ for a map M. In fact, if we define $P(M, \lambda) \equiv P(D(M), \lambda)$, we provide an automatic mechanism for converting propositions about $P(G, \lambda)$ for G planar into propositions about $P(M, \lambda)$, and vice versa. We shall feel free to state results for either maps or graphs.

Let us return to the question of the behavior of $P(G, \lambda)$ on the real line.

Theorem 6-14 If $\lambda < 0$ and $G \neq \Omega$, then for G connected

$$P(G, \lambda) = \begin{cases} > 0 & \text{if } n = \text{order of } G \text{ is even} \\ < 0 & \text{if } n = \text{order of } G \text{ is odd} \end{cases}$$

PROOF Suppose $P(G, \lambda) = \lambda^n - \alpha\lambda^{n-1} + \cdots + (-1)^{n-1}\alpha_{n-1}\lambda$. Then, for $\lambda < 0$, it is easy to check that $P(G, \lambda) = (-1)^n(|\lambda|^n + \alpha_1|\lambda|^{n-1} + \cdots + \alpha_n|\lambda|)$, where $|\lambda|$ denotes the absolute value of λ. The theorem now follows immediately.

With a bit more care, we can prove a stronger result due to Tutte (1969b). First, we need a lemma. Suppose $k(G)$ denotes the number of components of G.

Lemma 6-15 There is a polynomial $S(G, \lambda)$ such that $P(G, \lambda) = \lambda^{k(G)}S(G, \lambda)$.

PROOF Let $k = k(G)$ and let $G_1, \ldots, G_k$ be the components of G. Then

$$P(G, \lambda) = \prod_{i=1}^{k} P(G_i, \lambda)$$

by Theorem 6-7. By Lemma 6-12, each factor $P(G_i, \lambda)$ is divisible by λ and so $P(G, \lambda)$ is divisible by λ^k.

If G has a loop, we write $S(G, \lambda) = 0$.

Theorem 6-15 If G has no loops and $G \neq \Omega$, then $(-1)^{n+k}S(G, \lambda) > 0$, where n is the order of G and $k = k(G)$. Here $0 < \lambda < 1$.

PROOF Suppose the theorem is false and let G be a counter-example with a minimum number m of edges. Certainly, $m > 0$ since a graph H without edges has $n = k$ and $S(H, \lambda) = 1$. Also, G cannot be a *forest*. (A forest is a graph, each of whose components is a tree.) It is easy to compute that if G is a forest, then $P(G, \lambda) = \lambda^k(\lambda - 1)^{n-k}$ so $S(G, \lambda) = (\lambda - 1)^{n-k}$. Since $\lambda < 1$, $\lambda - 1 < 0$ so $(-1)^{n+k}(\lambda - 1)^{n-k} > 0$.

Since G is not a forest, it must have an edge A which is not a bridge. Hence, $k(G) = k(G'_A)$ and, for any edge B, $k(G) = k(G''_B)$.

Now, for any graph G of order n, let us write $T(G, \lambda) = (-1)^{n+k}S(G, \lambda)$ where $k = k(G)$. Since $P(G, \lambda) = P(G'_A, \lambda) - P(G''_A, \lambda)$, by Theorem 6-8 and $k(G) = k(G'_A) = k(G''_A)$, $S(G, \lambda) = S(G'_A, \lambda) - S(G''_A, \lambda)$. But G and G'_A have order n while G''_A has order $n - 1$. Hence,

$$T(G, \lambda) = T(G'_A, \lambda) + T(G''_A, \lambda)$$

Since G is loopless, G'_A is loopless with fewer edges than G so $T(G'_A, \lambda) > 0$ for $\lambda < 1$. Also, $T(G''_A, \lambda) > 0$ for $\lambda < 1$. Therefore, $T(G, \lambda) > 0$ for $\lambda < 1$.

This theorem implies our earlier result (Theorem 6-14) as well as the following interesting corollary.

Corollary 6-2 For $0 < \lambda < 1$, and G connected loopless of order n, $(-1)^n P(G, \lambda) < 0$.

PROOF Write $P(G, \lambda) = \lambda S(G, \lambda)$. Now $(-1)^{n+1} S(G, \lambda) > 0$ by Theorem 6-15. Hence,

$$(-1)^n P(G, \lambda) = (-1)^n \lambda S(G, \lambda) = (-\lambda)(-1)^{n+1} S(G, \lambda) < 0$$

Note that the hypothesis of connectivity is necessary in this corollary, for otherwise we could take $G = 2K_2$, the disjoint union of two copies of K_2. Then $P(G, \lambda) = \lambda^2 (\lambda - 1)^2$ so $P(G, \lambda) > 0$ and, in fact, $(-1)^4 P(G, \lambda) > 0$ for $0 < \lambda < 1$.

Let us put together Theorem 6-14 and Corollary 6-2 to obtain a picture of the behavior of $P(G, \lambda)$ for $\lambda \leq 1$ for G a loopless, connected, nontrivial graph. Assume for definiteness that the order n of G is even. Then we may graph $P(G, \lambda)$ as shown in Fig. 6-6.

Note that $P(G, \lambda)$ has a root at $\lambda = 0$, since $G \neq \Omega$ and at $\lambda = 1$ since G is connected and nontrivial and so must contain an edge.

One question which we have not yet asked is when a number r can be a root of some chromial. Of course, by Theorems 6-14 and 6-15 if $r < 0$ or $0 < r < 1$, then r is not a chromial root. Consider the positive root

$$\tau = \frac{1 + \sqrt{5}}{2} \text{ of the equation } x^2 = x + 1.$$

Theorem 6-16 The number $\tau + 1$ is never the root of a chromial.

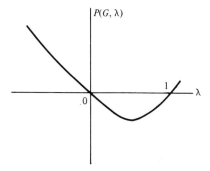

Figure 6-6

PROOF Suppose $(3 + \sqrt{5})/2 = \tau + 1$ is a root of $P(G, \lambda)$ for some loopless graph G. Then since $P(G, \lambda)$ is a polynomial with integral coefficients, $(3 - \sqrt{5})/2$ must also be a root. But $0 < (3 - \sqrt{5})/2 < 1$, contradicting Corollary 6-2.

The number τ is the classical "golden ratio" which the ancient Greeks believed expressed the ideal ratio of the length of a rectangle to its width. Although we have just shown that $\tau + 1$ is not itself the root of any chromatic polynomial, it has been observed empirically by Berman and Tutte (1969) that if G is a maximal planar graph, then $P(G, \lambda)$ tends to have a root near $\lambda = \tau + 1$. For this reason, we shall call $\tau + 1$ the *golden root*.

For example, Hall, Siry, and Vanderslice (1965) have computed the (complex) roots of the truncated icosahedron (i.e., of its chromatic polynomial). Of the four real (nonintegral) roots, one agrees with the golden root up to eight decimal places. In her thesis, Bari (1966) tabulated a number of chromatic polynomials. Berman and Tutte (1969) showed that most of these polynomials had a root agreeing with the golden root up to at least five decimal places.

In attempting to account for this remarkable behavior of chromials, Tutte (1970a) showed that $|P(G, 1 + \tau)| \leq \tau^{5-n}$ where M is a triangular map whose underlying graph G has n vertices. We shall present in full his proof of this rather deep result.

First some preliminaries. If $v \in V(G)$, we write G_v for $G-v$. We say that G is *wheel-like* at v if there is a circuit C of G satisfying: (a) $v \notin V(C)$, (b) each vertex of C is joined to v by at least one edge of G, and (c) no edge of G joins v to a vertex not belonging to C. We also say that C *encloses* v.

Theorem 6-17 Let G be a planar graph with vertex v. If v is enclosed by a circuit C with m vertices, then

$$P(G, 1 + \tau) = (-1)^m \tau^{1-m} P(G_v, 1 + \tau)$$

Before giving the proof, we shall need two lemmas whose proofs are exercises.

Lemma 6-16 Let $\eta = \tau + 1$. Then $\eta = \tau^2$, $\eta - 1 = \tau$, $\eta - 2 = \tau^{-1}$, $\eta - 3 = -\tau^{-2}$.

Define the *wheel with k spokes* W_k ($k \geq 3$) to be the graph obtained by joining a vertex v called the *hub* to each of the k vertices $v_1, \ldots, v_k$ of a circuit C_k by adding k new edges $[v, v_k]$, $1 \leq i \leq k$, called *spokes*.

Lemma 6-17 $P(W_k, \lambda) = \lambda(\lambda - 2)^k + (-1)^k(\lambda - 2))$.

Let us also state a useful corollary to Lemma 6-16 and Theorem 6-7.

Corollary 6-3 Let G be the union of two subgraphs H and K whose intersection is a complete graph K_m, $m = 0, 1, 2,$ or 3. Then

$$P(G, \eta) = \tau^{-s} P(H, \eta) P(K, \eta)$$

where $s = 0, 2, 3,$ or 2, respectively.

Now we can prove the theorem.

PROOF OF THEOREM 6-17 Let n be the order of G and m the number of edges. We induct on $k = n + m$, the theorem being vacuously true for $k = 0$. Suppose inductively that it holds whenever $k < q$ $(q > 0)$ and let $k = q$.

Case (1) Suppose there is a vertex $x \neq v$ which is not in C.

Subcase (α) The vertex x is isolated (of degree 0). Since $G = x \cup G_x$ and v is enclosed by C in G_x and since $G_v = x \cup (G_x)_v$ we have

$$\begin{aligned}
P(G, \eta) &= P(x, \eta) P(G_x, \eta) \\
&= (-1)^m \tau^{1-m} P(x, \eta) P((G_x)_v, \eta) \\
&= (-1)^m \tau^{1-m} P(G_v, \eta)
\end{aligned}$$

Subcase (β) The vertex x is joined to another vertex y (necessarily not equal to v) by at least two distinct edges A and B. Let $H = G'_B$. Then $H_v = (G_v)'_B$. Since H has fewer edges than G, the theorem holds for H so we have

$$\begin{aligned}
P(G, \eta) &= P(H, \eta) \\
&= (-1)^m \tau^{1-m} P(H_v, \eta) \\
&= (-1)^m \tau^{1-m} P((G_v)'_B, \eta) \\
&= (-1)^m \tau^{1-m} P(G_v, \eta)
\end{aligned}$$

Subcase (γ) x is joined to $y \neq v$ by a unique edge A. This case is similar to the preceding one and is left as an exercise.

Case (2) Suppose that v is the only vertex of G not in C. Enumerate the vertices of C, $v_1, \ldots, v_m$ in their cyclic order. If $m = 1$, the theorem is true since G and G_v contain a loop. Hence $m \geq 2$.

Subcase (α) There is an edge A of G whose endpoints are nonconsecutive vertices of C, say v_1 and v_j, where $2 < j < m$. There are edges A_1 and A_j joining v to v_1 and v_j, respectively. The edges A, A_1, and A_j define a complete subgraph L in G. Since G is planar, it follows that G is the union of proper subgraphs H and K which intersect in L, where $v_2 \in V(H)$ and $v_{j+1} \in V(K)$. The inductive hypothesis applies to H and K. Moreover, v is enclosed in H by a circuit of j edges and in K by a circuit of $m - j + 2$ edges. Also, observe that $G_v = H_v \cup K_v$ and $H_v \cap K_v$ is the complete graph determined by A. Hence, since $\eta(\eta - 1)(\eta - 2) = \tau^2$,

$$P(G, \eta) = \tau^{-2} P(H, \eta) P(K, \eta)$$

$$= \tau^{-2}[(-1)^j \tau^{1-j} P(H_v, \eta)][(-1)^{m-j+2} \tau^{-1-m+j} P(K_v, \eta)]$$
$$= (-1)^m \tau^{-2-m} P(H_v, \eta) P(K_v, \eta)$$
$$= (-1)^m \tau^{1-m} P(G_v, \eta)$$

Subcase (β) G is a wheel with m spokes. This is an easy exercise based on Lemma 6-17.

The proof of Theorem 6-17 is now complete.

Now we want to prove Tutte's main result. Let $Z(n)$ denote the class of all triangular maps M with n vertices and let $Z(n, m)$ denote the class of all maps M with n vertices with exactly one m-gon and all other regions triangles.

Note that if $G = U(M)$ is simple, then $M \in Z(n)$ if and only if G is maximal planar. However, if G is not simple, then the way in which G is drawn in the plane determines whether or not the resultant map $M \in Z(n)$. If $M \in Z(n, m)$ and $G = U(M)$ is simple, then an easy argument based on Euler's formula shows that G must have exactly $3n - 6 - (m - 3)$ edges. Thus, if $m = 5$, G has $3n - 8$ edges. On the other hand, if M has 2 four-sided regions and all others triangular, then G would still have $3n - 8$ edges. For these reasons we must state the hypotheses of the next theorem in terms of M, rather than $U(M)$, even though the conclusion involves the chromatic polynomial of $U(M)$.

Theorem 6-18 (*a*) If $M \in Z(k)$, then

$$|P(U(M), \eta)| \le \tau^{5-k}$$

(*b*) If $M \in Z(k, m)$, where $2 \le m \le 5$, then

$$|P(U(M), \eta)| \le \tau^{3+m-k}$$

PROOF Let $G = U(M)$. The theorem is vacuously true for $k = 0$ since $Z(0) = \varnothing = Z(0, m)$. Assume inductively that the theorem is valid whenever $k < q$ ($q > 0$) and suppose $k = q$.

Case (1) $M \in Z(q)$.

Subcase (α) G has a circuit C of length 2 (that is, G has a pair of parallel edges). Since M has no 2-gons, C must separate M into maps M_1 and M_2 in $Z(q_1, 2)$ and $Z(q_2, 2)$, respectively, where $0 < q_i < q$ ($i = 1, 2$) and $q_1 + q_2 = q + 2$. Let $H_i = U(G_i)$ ($i = 1, 2$). Then $H_1 \cup H_2 = G$ and $H_1 \cap H_2 = C$. Let A and B denote the two edges of C. If K is any subgraph of G containing both A and B, we write K' for K'_A. Now $P(K, \lambda) = P(K', \lambda)$ for such K, and C' is a complete graph of order 2. Hence, by Corollary 6-3 (to Lemma 6-17) and the preceding remarks, we have

$$|P(G, \eta)| = |P(G', \eta)| = \tau^{-3} |P(G'_1, \eta)| \, |P(G'_2, \eta)|$$
$$= (\tau^{-3})(\tau^{5-q_1})(\tau^{5-q_2})$$
$$= \tau^{5-q}$$

Subcase (β) G has no circuit of length 2. Since M is a triangulation, M has some vertex v of degree at most 5. If $G = K_3$, then $P(G, \eta) = \tau^2 = \tau^{5-q}$. Hence, G is wheel-like at v with v enclosed in a circuit C with m edges, where m is the degree of v. Let M_v be the map obtained from M by deleting v and its incident edges. Thus, $U(M_v) = G_v$ and $M_v \in Z(q - 1, m)$. Hence, by Theorem 6-17 and the inductive hypothesis,

$$\begin{aligned}
\left| P(G, \eta) \right| &= \tau^{1-m} \left| P(M_v, \eta) \right| \\
&\leq \tau^{1-m} \tau^{3+m-(q-1)} \\
&= \tau^{5-q}
\end{aligned}$$

Case (2) $M \in Z(q, m)$, where $m = 2, 3, 4,$ or 5.

Subcase (α) $m = 2$. Since M has exactly one 2-gon, deleting one of the edges of the 2-gon produces a triangular map $N \in Z(q)$. Thus, by case (1),

$$\begin{aligned}
\left| P(G, \eta) \right| &= \left| P(U(N), \eta) \right| \\
&\leq \tau^{5-q} \\
&= \tau^{3+m-q}
\end{aligned}$$

Subcase (β) $m = 3$. Hence, $M \in Z(q)$, so by case (1)

$$\left| P(G, \eta) \right| \leq \tau^{5-q} < \tau^{3+m-q}$$

Subcase (γ) $m = 4$ or 5. Let $v_1, \ldots, v_m$ be an enumeration of the vertices of C in clockwise order. By planarity, we can find two vertices, say v_1 and v_3, in the boundary of the m-gon such that no edge of G joins v_1 and v_3, for if some edge does join v_1 and v_3 no edge joins v_2 and v_m. Add a new edge A passing through the interior of the m-gon and joining v_1 and v_3; call the resulting map N and set $J = U(N)$. Note $N \in Z(q, m-1)$. Let N_1 be the map obtained from N by contracting A to a new vertex x and deleting either one of the edges, say B, joining x to v_2. Then, $N_1 \in Z(q-1, m-2)$. If $H = U(N_1)$, then clearly $G = J'_A$ and $H = (J'_A)'_B$. Therefore, by Theorem 6-8,

$$P(G, \lambda) = P(J, \lambda) + P(H, \lambda)$$

Since for any two real numbers a and b, $\left| a + b \right| \leq \left| a \right| + \left| b \right|$, we have

$$\left| P(G, \lambda) \right| \leq \left| P(J, \lambda) \right| + \left| P(H, \lambda) \right|$$

If $m = 4$, $N \in Z(q, 3) = Z(q)$ and $N_1 \in Z(q - 1, 2)$. Hence, by the inductive hypotheses and case (1),

$$\begin{aligned}
\left| P(G, \eta) \right| &\leq \tau^{5-q} + \tau^{5-(q-1)} \\
&= \tau^{5-q}(1 + \tau) \\
&= \tau^{7-q} \\
&= \tau^{3+m-q}
\end{aligned}$$

If $m = 5$, $N \in Z(q, 4)$ and $N_1 \in Z(q - 1)$. Therefore, by induction and the preceding paragraph,

$$
\begin{aligned}
\left| P(G, \eta) \right| &\leq \tau^{7-q} + \tau^{5-(q-1)} \\
&= \tau^{8-q} \\
&= \tau^{3+m-q}
\end{aligned}
$$

This completes the proof of Theorem 6-18.

This theorem has a rather obvious corollary which is nevertheless worth stating.

Corollary 6-4 Let G_i be any sequence of maximal planar graphs, where G_i has order n_i and $\lim\limits_{i \to \infty} n_i = \infty$. Then

$$
\lim_{i \to \infty} P(G_i, 1 + \tau) = 0
$$

Thus, the golden root $1 + \tau$, while not the root of any graph, is, in a sense, arbitrarily close to being a root of all sufficiently large maximal planar graphs. We shall return to this interesting situation in Sec. 6-8.

6-6 THE COEFFICIENTS OF A CHROMIAL

In the general study of chromials $P(G, \lambda)$, where G is any graph (not necessarily planar), we want to interpret the coefficients $\alpha_i(G)$ in terms of the graph G.

First, however, we shall describe, for t an integer, $P(G, t)$ in terms of the "factorial powers" $t_{(r)} = t(t - 1) \cdots (t - r + 1) = t!/(t - r)!$, where $1 \leq r \leq t$. If r is a positive integer, let $E(G, r)$ denote the number of ways of coloring G using exactly r colors where two colorings are regarded as the same if they agree up to a permutation of the colors. Thus, $E(G, r)$ is the number of ways in which the vertices of G can be partitioned into r disjoint nonempty independent sets.

Lemma 6-18 If $G \neq \Omega$, then $P(G, t) = \sum\limits_{r=1}^{t} \binom{t}{r} r! E(G, r)$.

PROOF Each coloring of G from t colors uses some subset of r colors from t. There are $\binom{t}{r}$ such subsets and $r!$ different permutations of the colors.

Theorem 6-19 If $G \neq \Omega$, then $P(G, t) = \sum\limits_{r=1}^{t} t_{(r)} E(G, r)$.

PROOF $\binom{t}{r} r! = t_{(r)}$.

Thus, $P(G, t)$ can be expressed as a "polynomial" in the $t_{(r)}$ with coefficients $E(G, r)$. A corresponding interpretation for the coefficients $\alpha_i(G)$ was given by Whitney (1932b).

Theorem 6-20 Let G be a graph of order n. Then

$$\alpha_{n-p}(G) = \sum_{r=0}^{k} (-1)^{n-p+r} N(p, r)$$

where $N(p, r)$ denotes the number of spanning subgraphs of G with p components and r edges, and k is the number of edges in G.

Before proving the theorem, let us note that if $p = n - 1$, then $r = 1$; that is, the only spanning subgraphs of G with $n - 1$ components are those with exactly one edge. Hence, $(-1)^{n-p+r} = (-1)^2 = 1$ and $N(p, r) = N(n - 1, 1) = k$. Thus, $\alpha_1(G) = k$, which agrees with our earlier result (Lemma 6-13).

PROOF OF THEOREM 6-20 We shall proceed with a counting argument. Let t be a fixed positive integer. A *t-assignment*, or simply, an *assignment*, is any function f from $V(G)$ to $\{1, \ldots, t\}$. An assignment f is a coloring if $f(v) \neq f(w)$ whenever v is adjacent to w. Call f *highly improper* or *hi* if $f(v) = f(w)$ whenever v is adjacent to w. Given a subgraph H of G and an assignment f of G, call H hi if f is hi regarded as an assignment of H. Any assignment f of G has a maximal hi subgraph $H = H(f)$ obtained by removing all edges in G joining any pair of vertices v and w for which $f(v) \neq f(w)$. If f is a coloring, H is edgeless. If $H(f)$ has p components, then each component is associated with a color, so there are t^p different assignments with the same hi subgraph.

First subtract from the total t^n of all assignments those assignments f for which $H(f)$ has only one edge. If we subtract $\Sigma_p N(p, 1) t^p$, we shall have subtracted these but much more besides. Suppose A and B are two disjoint edges in G. The contribution from the subgraph determined by A includes assignments in which the two endpoints of B receive the same color which are therefore hi assignments for the subgraph consisting of A and B. Moreover, it will have been subtracted twice, once for A and once for B.

To restore the balance, we can add $\Sigma_p N(p, 2) t^p$. This will compensate for the double subtraction of the assignments producing two-edge subgraphs, but will now necessitate a corresponding compensation for the three-edge subgraphs. Thus, we obtain

$$P(G, t) = t^n - \sum_p N(p, 1) t^p + \sum_p N(p, 2) t^p - \cdots$$

We can rewrite this as

$$P(G, t) = \sum_{r=0}^{k} \sum_{p=1}^{n} (-1)^r N(p, r) t^p$$

$$= \sum_{p=1}^{n} \sum_{r=0}^{k} (-1)^r N(p, r) t^p$$

Thus, the coefficient of t^p is $\sum_{r=0}^{k} (-1)^r N(p, r)$. But by (*) (see page 143) the coefficient of t^p is $(-1)^{n-p} \alpha_{n-p}$. Hence,

$$\alpha_{n-p} = \sum_{r=0}^{k} (-1)^{n-p+r} N(p, r)$$

This completes the proof of Theorem 6-20.

The theorem would require examining all 2^k spanning subgraphs of G. However, Whitney (1932b) showed that it is possible to restrict one's attention to a much smaller class. List the edges of G in some arbitrary order and remove from each circuit of G the highest numbered edge. Call the remaining subgraphs *broken circuits*.

Theorem 6-21 Theorem 6-20 holds if, in computing $N(p, r)$, we consider only those subgraphs of G which do not contain any broken circuit.

PROOF The idea of the proof is to show that those subgraphs which contain a broken circuit can be paired off so that their contributions to $P(G, t)$ cancel.

Let B be a broken circuit of G whose missing edge is b. If $H \subseteq G$ contains B but not b, form H^* from H by adding the edge b. The subgraphs H and H^* have the same number of components, but H^* has one more edge than H. Hence, their contributions cancel.

The difficult part of the proof is to show that all of the subgraphs which contain some broken circuit can be paired so that their contributions nullify each other. It then follows that only subgraphs not containing broken circuits need be considered.

Read (1968) asked whether it is true that, for all chromatic polynomials $P(G, \lambda)$, the $\alpha_i(G)$ first increase and then decrease. Chvátal (1970a) has shown that a similar phenomenon does hold for some of the coefficients $E(G, r)$ of $t_{(r)}$ in the expansion of $P(G, t)$ as a "polynomial" in $t_{(r)}$ given in Theorem 6-19.

Theorem 6-22 Let G be a graph of order $n > 0$ and set $\beta_i = E(G, i)$, $1 \le i \le n$. If $(j + 2)^{j-1} \le 2^n$, then $\beta_1 \le \beta_2 \le \cdots \le \beta_j$.

PROOF Given any partition π of $V(G)$ into $j - 1$ disjoint independent subsets, one can "grow" new partitions into j classes by splitting one of the $j - 1$ subsets into two new classes. If the original classes had $n_1, n_2, \ldots, n_{j-1}$

elements, then the number of the new partitions which can be grown from π is

$$\sum_{i=1}^{j-1} \frac{1}{2}(2^{n_i} - 2) \geq \frac{j-1}{2}(2^{n/(j-1)} - 2)$$

Running over all possible π we grow the same partition π' into j classes at most $\binom{j}{2}$ times since, having grown π' from π, we can reconstruct π from π' by amalgamating two of the j classes; this can be done in $\binom{j}{2}$ ways. Thus, we have

$$\beta_j \geq \frac{1}{\binom{j}{2}} \beta_{j-1} \frac{j-1}{2}(2^{n/(j-1)} - 2) = \beta_{j-1} \frac{2^{n/(j-1)} - 2}{j}$$

Thus, if $2^{n/(j-1)} \geq j + 2$, then $\beta_j \geq \beta_{j-1}$.

6-7 THE BIRKHOFF–LEWIS CONJECTURE; RELATIVE AND UNIQUE COLORINGS

If M is a map, recall that $P(M, \lambda)$ means $P(D(M), \lambda)$. If M is cubic, then $\chi(M) \geq 3$ so $P(M, \lambda)$ is divisible by $\lambda(\lambda - 1)(\lambda - 2)$. Hence, $[P(M, \lambda)]/[\lambda(\lambda - 1)(\lambda - 2)]$ is a polynomial in λ. Define for polynomials f and g, $f(x) \ll g(x)$ if the coefficients of $f(x)$ are nonnegative and not greater than the corresponding coefficients of $g(x)$. Now define $S(y) \ll T(y)$ for $y \geq c$ for polynomials S and T if, setting $x = y - c$, $f(x) = S(y)$, and $g(x) = T(y)$, the relation $f(x) \ll g(x)$ holds.

Birkhoff and Lewis (1946) used these notions to introduce a strong form of the 4CC.

Conjecture BL If M is any cubic map with $n + 3$ regions, then

$$(\lambda - 3)^n \ll \frac{P(M, \lambda)}{\lambda(\lambda - 1)(\lambda - 2)} \ll (\lambda - 2)^n \quad \text{for} \quad \lambda \geq 4$$

Note that if $P(M, \lambda)$ satisfies the Birkhoff–Lewis conjecture, then $P(M, 4) \geq 4 \times 3 \times 2 \times (4 - 3)^n = 24$. Thus, if Conjecture BL holds, every cubic map (or dually every maximal planar graph) has at least 24 four-colorings.

Birkhoff and Lewis showed that if $\lambda \geq 4$ is replaced by $\lambda \geq 5$, then Conjecture BL is true. Moreover, they verified Conjecture BL for $n \leq 8$.

Let us set $x = \lambda - 4$ and define the R-polynomial of a cubic map M by

$$R(M, x) = \frac{P(M, \lambda)}{\lambda(\lambda - 1)(\lambda - 2)}$$

Thus, Conjecture BL is equivalent to the following conjecture.

Conjecture BL′ Let M be a cubic map with $n + 3$ regions. Then $(x + 1)^n \ll R(M, x) \ll (x + 2)^n$.

A five-regular map M is characterized by the following properties: (1) M is cubic, (2) $D(M)$ is five-connected, and (3) if Z is any cycle of length 5 in $D(M)$ whose removal separates the graph into components G_1 and G_2, then either G_1 or G_2 consists of a single vertex. We shall see later that to prove the 4CC, it suffices to prove that every five-regular map can be four-colored.

Bari (1969) has shown that for the class of five-regular maps BL′ does hold for the first four coefficients of $R(M, x)$ by computing these first four coefficients.

Theorem 6-23 If M has $n + 3$ regions, then

$$R(M, X) = x^n + nx^{n-1} + \binom{n-1}{2}(n+2)x^{n-2} + \binom{n-2}{6}(n^2 + 5n - 18)x^{n-3} + \cdots$$

Recent work of Albertson and Wilf (1974a and b) can also be used to support the Birkhoff–Lewis conjecture. Let M be a map whose outer boundary $B(M)$ is a cycle. Any four-coloring of $G = U(M)$ induces a four-coloring of $B(M)$ and the set of all such four-colorings of $B(M)$ is called the set of *admissible boundary colorings*. Let Ψ denote the cardinality of this set; Ψ depends on M, not just on G. Clearly, the 4CC holds if and only if $\Psi > 0$. Remember that Ψ counts as distinct two-colorings which differ by a nonidentity color permutation. Albertson and Wilf have asserted the following.

Conjecture AW Let M be a map whose boundary $B(M)$ is a circuit of length k. Then $\Psi \geq 3 \times 2^k$.

This conjecture certainly implies the 4CC. Moreover, they were able to show the following theorem.

Theorem 6-24 If the 4CC holds, then Conjecture AW is valid for $k = 3, 4, 5$, and 6.

Their work has an interesting consequence.

Corollary 6-5 If the 4CC holds, then every maximal planar graph has at least 24 four-colorings.

PROOF $\Psi \geq 3 \times 2^3 = 24$.

Note, for example, that K_3 has 24 different four-colorings since there are 4 distinct three-element subsets of $\{1, \ldots, 4\}$ and 6 distinct permutations of any three-element set. Thus, we can state another equivalent version of the 4CC.

Conjecture C_{29} Every maximal planar graph has at least 24 four-colorings.

The notion of admissible boundary colorings is a special case of the more general notion of relative colorings (see Kainen, 1973a). Let us call (G, H) a *pair* if H is a full subgraph of G. (Note that if $H = \Omega$, then H is always full in G.) If c is an r-coloring of G in which all r colors are used, we write $|c| = r$.

If (G, H) is a pair of graphs, c is a coloring of G, and d a coloring of H, we say that c *extends* d if $c \mid H = d$. The *relative chromatic number* $\chi(G, H)$ of the pair (G, H) is defined by

$$\chi(G, H) = \sup_{d} \inf_{c} |c| - |d|$$

where d is any coloring of H and c is any coloring of G extending d. Thus, $\chi(G, H)$ is the minimum number of *new* colors which will be needed to extend any coloring of H to a coloring of G.

Consider a *triple* of graphs (G, H, K), where K is full in H and H is full in G. If c is an r-coloring of K, then c can be extended to d, an s-coloring of H, where $s \leq r + \chi(H, K)$. Now d can be extended to e, a t-coloring of G, where $t \leq s + \chi(G, H)$. Therefore, $t \leq r + \chi(G, H) + \chi(H, K)$ and hence we have proved the following lemma.

Lemma 6-19 For any triple (G, H, K), $\chi(G, K) \leq \chi(G, H) + \chi(H, K)$.

Corollary 6-6 For any pair (G, H), $\chi(G) \leq \chi(G, H) + \chi(H)$.

PROOF Take $K = \Omega$ in Lemma 6-19 and observe that $\chi(H) = \chi(H, \Omega)$. The next lemma is trivial.

Lemma 6-20 For any pair (G, H), $\chi(G, H) \leq \chi(G - H)$.

Corollary 6-7 For any pair (G, H), $\chi(G) - \chi(H) \leq \chi(G, H) \leq \chi(G - H)$.

Now suppose that G is planar. Then $G - H$ is planar, so by the five-color theorem $\chi(G, H) \leq 5$ for any H. Conversely, if $\chi(G, H) \leq 5$ for all H, then in particular $\chi(G) = \chi(G, \varnothing) \leq 5$. This suggests the following.

Conjecture C_{30} Let G be planar. Then $\chi(G, H) \leq 4$ for all pairs (G, H).

Theorem 6-25 Conjectures C_{26} and C_{30} are equivalent.

PROOF If C_{30} holds, then $\chi(G) = \chi(G, \varnothing) \leq 4$ for any planar G. Conversely, if C_{26} holds, then for any planar G and full subgraph H, $\chi(G, H) \leq \chi(G - H) \leq 4$ since $G - H$ is planar.

Suppose now that G is planar and that H is a connected (nonempty) full subgraph of G. One then finds oneself unable to construct an example in which even four new colors are needed to extend some coloring of H to a coloring of G; in fact, three always seem to suffice. Hence, a new conjecture.

Conjecture C_{31} Let G be planar. Then $\chi(G, H) \leq 3$ for all pairs (G, H) in which H is connected (and nonempty).

Theorem 6-26 Conjectures C_{26} and C_{31} are equivalent.

PROOF One-half of the theorem is trivial. Suppose that G is any planar graph and let H be a single vertex. By Corollary 6-6 (to Lemma 6-19), $\chi(G) \leq \chi(G, H) + \chi(H) \leq 3 + 1 = 4$.

Conversely, suppose that G is any planar graph and that H is a connected full subgraph. Let c be an r-coloring of H with colors $c_1, \ldots, c_r$. We must extend c to a coloring d of G using at most three new colors.

Since H is connected, we may find a spanning tree T in H. Now shrink T to a single point x and let $\bar{G}$ be the corresponding graph. Specifically, $V(\bar{G}) = V(G) - V(H) \cup \{x\}$. Two vertices other than x are adjacent in $\bar{G}$ if and only if they were adjacent in G. A vertex v is adjacent to x if and only if v was adjacent in G to some vertex w in $V(H)$. Finally, we delete all loops and parallel edges. Note that $\bar{G}$ is still planar since we have collapsed a contractible subgraph.

By Conjecture C_{26} $\chi(\bar{G}) \leq 4$ so we can four-color $\bar{G}$ by a coloring e. Moreover, we can assume that the color $e(x)$ which e assigns to x is one of the original r colors $c_1, \ldots, c_r$, say c_1, while the other three colors are all new. We now define d as follows:

$$d(v) = \begin{cases} c(v) & \text{if } v \in V(H) \\ e(v) & \text{if } v \in V(G) - V(H) \end{cases}$$

Obviously, the only thing which needs checking is that if $v \in V(G) - V(H)$, $w \in V(H)$, and $[v, w] \in E(G)$, then $d(v) \neq d(w)$. Suppose $d(v) = d(w)$. Then since $d(v) = e(v)$, we must have $d(v) = e(v) = c_1$. But $[v, w] \in E(G)$ means v is adjacent to x in $\bar{G}$ and hence $e(v) \neq e(x) = c_1$.

This completes the proof of Theorem 6-26. A different proof has been given by Levow (1973). Moreover, Bernhart (1973a) has shown, without assuming the 4CC, that Conjecture C_{31} does hold whenever H is nontrivial.

Let us return for a moment to $P(G, t)$, t a positive integer. We know that $\chi(G)$ is the smallest positive integer t for which $P(G, t) \neq 0$. What can we say about $P(G, \chi(G))$? In general, not much. However, if we assume that

$$P(G, k) = k!$$

where $k = \chi(G)$, we obtain the definition of a *uniquely k-colorable* (or *uniquely colorable*) graph. Thus, G is uniquely k-colorable if and only if there is a unique partition of $V(G)$ into k disjoint nonempty subsets $V_1, \ldots, V_k$. (We call these subsets V_i *color classes.*)

Lemma 6-21 Let G be uniquely k-colorable with partition $V(G) = V_1 \cup \cdots \cup V_k$. If $v \in V_i$, then for each $j \neq i$, $1 \leq j \leq k$, there is a vertex $v_j \in V_j$ adjacent to v.

PROOF If not, we could recolor v.

Corollary 6-8 If G is uniquely k-colorable, then $\delta(G) \geq k - 1$.

Another necessary condition was found by Cartwright and Harary (1968).

Theorem 6-27 Let G be uniquely k-colorable with color classes $V_1, \ldots, V_k$. Then the union of any two of the color classes induces a connected subgraph of G.

PROOF Suppose two color classes V_1 and V_2 induce a disconnected subgraph $H = G(V_1 \cup V_2)$. Let H_1 and H_2 be distinct components of H. By Lemma 6-21, both H_1 and H_2 contain vertices of V_1 and V_2. Interchanging the colors of the vertices in $V_1 \cap H_1$ and $V_2 \cap H_1$ produces a different partitioning of G into color classes, which is impossible.

Of course, an immediate consequence of the theorem is that uniquely k-colorable graphs ($k \geq 2$) are connected. However, we can give a stronger result due to Chartrand and Geller (1969).

Theorem 6-28 Every uniquely k-colorable graph G is $(k - 1)$ connected.

PROOF Assume without loss of generality that G is not complete. If G is not $(k - 1)$-connected, there exists a set U of $k - 2$ vertices such that G–U is not connected. Hence, there are at least two distinct colors, say C_1 and C_2, not assigned to any point of U. By Theorem 6-27 the set of all vertices in G colored C_1 or C_2 is contained within the same component G_1 of G–U. Now color any vertex of G–U not in G_1 with either C_1 or C_2 to obtain a contradiction.

In addition, Chartrand and Geller proved the following theorem.

Theorem 6-29 Every uniquely four-colorable planar graph is maximal planar.

A result due to Hedetneimi (see Chartrand and Geller, 1969) shows that regardless of the truth or falsity of the 4CC, there are no uniquely five-colorable planar graphs. We give a somewhat different proof.

Theorem 6-30 No planar graph G is uniquely five-colorable.

PROOF Suppose G is planar and uniquely five-colorable. Choose two distinct vertices v, w in one of the color classes (otherwise G would be K_5) and add an edge $e = [v, w]$ to G. It was proved in Kainen (1974a) that $\chi(G + e) \leq 5$. But since G is uniquely five-colorable, we would need six colors to color $G + e$.

Let us close with the notion of semi-uniquely k-colorable graphs due to Greenwell (1971). If $\chi(G) = k$, G is semi-uniquely k-colorable if there exist v, $w \in V(G)$ such that every partition of $V(G)$ into color classes places v and w into the same class. Call v and w *linked*.

Conjecture C_{32} If G is a semi-uniquely four-colorable planar graph in which v and w are linked, then $G + [v, w]$ is not planar.

Theorem 6-31 Conjectures C_{32} and C_{26} are equivalent.

PROOF If $G + [v, w]$ is planar, then $\chi(G + [v, w]) = 5$. Hence, Conjecture C_{26} implies Conjecture C_{32}. Conversely, suppose C_{32} holds. We prove C_{26} by induction on the number m of lines in a planar graph G. If $m = 0$, $\chi(G) = 1 < 4$. Suppose C_{26} holds whenever $m < q$ $(q > 0)$ and let $m = q$. Choose an edge $e \in E(G)$, $e = [v, w]$. By induction, $\chi(G - e) \leq 4$. If v and w were linked in $G - e$, then, by Conjecture C_{32}, $G - e + e = G$ would be nonplanar. Hence, there is a four-coloring of $G - e$ which assigns different colors to v and w; this four-coloring is a four-coloring of G.

Using the same method as in the proof of Theorem 6-30, one can establish that no planar graph is semi-uniquely five-colorable.

6-8 THE BERAHA NUMBERS

In Sec. 6-4, we considered the result of evaluating, for G a triangulation, chromials $P(G, \lambda)$ at $\lambda = 1 + \tau$, where $\tau = (1 + \sqrt{5})/2$. Beraha (1974) has suggested the examination of $P(G, \lambda)$ at the points $\lambda = B(n)$, where $B(n) = 2 + 2\cos(2\pi/n)$. The numbers $B(n)$ are now called *Beraha numbers*; the first few are given in Table 6-1.

Table 6-1

n	$B(n)$
1	4
2	0
3	1
4	2
5	$\tau + 1 = 2.618\,033\,988\,7$, approximately
6	3
7	$\sigma = 3.246\,98$, approximately
8	$2 + \sqrt{2} = 3.414\,21$, approximately
9	$2 + 2\cos(2\pi/9) = 3.532\,088$, approximately
10	$\tau + 2$

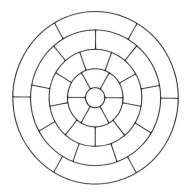

Figure 6-7

We shall explain the significance of some of these values shortly. First, however, let us consider the chromatic polynomial of the truncated icosahedron (see Fig. 6-7) which is obtained by truncating the regular icosahedron at each of its 12 vertices to produce a map M with 12 pentagons and 20 hexagons. Furthermore, no two pentagons in M are adjacent. Hall, Siry, and Vanderslice (1965) evaluated $P(M, \lambda)$ and found exactly four nonintegral real roots whose approximately numerical values are given in Table 6-2. Note the good agreement between n_1 and $\tau + 1$, n_2 and σ, n_3 and $B(8)$, and n_4 and $B(9)$.

Let us see another way in which the golden root $1 + \tau$ intrudes in this area. Suppose that G is any planar graph containing a quadrilateral face F. Let $A_1, \ldots, A_4$ in Fig. 6-8 denote, respectively, the number of ways in which G can be colored from n colors (n a positive integer) so that the four vertices of the bounding cycle C of F receive colors as indicated. Our approach follows closely that of Hall (1973). Thus A_1 counts the number of ways G can be colored from n colors with four colors appearing in C; A_2 and A_3 count the number of ways with three colors appearing; and A_4 counts the ways with two colors. The A_i, regarded as polynomials in n, are called *constrained chromials*.

Consider the pictures of Fig. 6-9 which represent G with C modified as

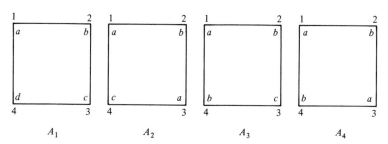

Figure 6-8

Table 6-2

$n_1 = 2.618\,033\,99$
$n_2 = 3.246\,991\,9$
$n_3 = 3.415\,399\,30$
$n_4 = 3.520\,045\,93$

indicated. If we fail to distinguish between the graph and its chromatic polynomial, we have

$$Z_1 = A_1 + A_3$$
$$Z_2 = A_1 + A_2$$
$$Y_1 = A_2 + A_4$$
$$Y_2 = A_3 + A_4$$

That is, we have solved for the *free* chromials in terms of the constrained chromials. If the process is reversed, one finds, for example, that

$$(n^2 - 3n + 1)A_4 = -Z_1 + (n - 2)Y_2 + (n - 3)Y_1$$

(This was first discovered by Birkhoff and Lewis (1948); see also Hall (1971) for a shorter proof based on a method of Tutte.) The point we want to make is that the positive root of the coefficient of A_4 in the above formula is exactly $\tau + 1$.

Analysis of five-cycles gives no new information, but in their study of the six-ring, Hall and Lewis (1946) derive equations for constrained chromials in terms of free chromials. The factors involved are $n^2 - 3n + 1$ and a new polynomial $n^3 - 5n^2 + 6n - 1$ whose largest root is none other than $\sigma = B(7)$. The root σ is called the *silver root*.

For the seven-cycle, the number $B(8) = 2 + \sqrt{2}$ appears, and Tutte (1971) conjectures that this phenomenon continues, with $B(n + 1)$ associated to an analysis of the n-ring.

Another of the Beraha numbers figures in a recent result of Tutte (1969b).

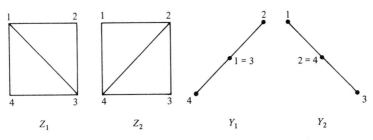

$Z_1 \qquad\qquad Z_2 \qquad\qquad Y_1 \qquad\qquad Y_2$

Figure 6-9

Theorem 6-32 If G is a planar triangulation with k vertices, then $P(G, \tau + 2) = (\tau + 2)\tau^{3k-10}P^2(G, \tau + 1)$.

Of course, since $P(G, \tau + 1) \neq 0$, we conclude the following.

Corollary 6-9 If G is any planar triangulation, then $P(G, \tau + 2) > 0$.

In his work, Tutte (1972b) appears to have fully vindicated Beraha's guess that the numbers $B(n)$ are important in the theory of chromatic polynomials by obtaining equations when $\lambda = B(n)$, $n \geq 5$, for $l(y, z, \lambda)$, which represents a certain sum of chromials (see Tutte, 1972a).

Since the Beraha numbers converge to 4, it is possible that by studying the behavior of $P(G, \lambda)$ as $\lambda = B(5)$, $B(6), \ldots$, we may hope to obtain information on $P(G, 4)$.

PROJECTIVE GEOMETRY AND CHAIN GROUPS

7-1 INTRODUCTION

A very strong form of the 4CC is known as Hadwiger's conjecture. This is actually a sequence of conjectures, one for each value of $k \geq 1$. We discuss this conjecture in Sec. 7-2. Then in Sec. 7-3 we give a brief account of Whitney's theory of matroids and its application by Tutte to an elegant conjecture which relates Hadwiger's conjecture to Tutte's earlier conjecture of Sec. 4-2. The last section of this chapter also deals with work of Tutte, this time on finite geometries and the 4CC.

7-2 HADWIGER'S CONJECTURE

An *edge contraction* of a graph G is obtained by removing two adjacent vertices u and v and adding a new vertex, w, adjacent to those vertices to which u or v was adjacent. A graph G is *contractible* to a graph H if H can be obtained from G by a sequence of edge contractions. We shall also call H a *contraction* of G. Note that G is contractible to H if and only if there is a connected homomorphism (see Ore, 1962, p. 85) from G onto H; that is, a function from $V(G)$ onto $V(H)$ such that the inverse image of each vertex in $V(H)$ induces a connected subgraph of G. The following conjecture is due to Hadwiger (1943).

Conjecture H Every connected graph G with $\chi(G) \geq r$ is contractible to a complete graph on r vertices.

The truth of this conjecture for $r < 5$ has been verified by Dirac (1952a).

Theorem 7-1 For $r < 5$, every connected graph G with $\chi(G) \geq r$ is contractible to K_r.

The proof for $r = 4$ is involved, but for $r < 4$ it is easy. If $r = 1$, there is a unique contraction to K_r. If $r = 2$, G has at least two vertices. But for any connected graph with ≥ 2 vertices, there are many contractions to K_2. Finally, suppose $r = 3$. If G is not contractible to K_3, then G is a tree and so $\chi(G) = 2$. Hence, G is contractible to K_3.

Hadwiger's conjecture, in the case $r = 5$, turns out to be particularly significant.

Conjecture C$_{33}$ Let $\chi(G) \geq 5$. Then G is contractible to K_5.

This conjecture certainly implies C_1. Suppose $\chi(G) = 5$. Then G is contractible to K_5. But K_5 is nonplanar so G is nonplanar since any contraction of a planar graph is planar. Therefore, C_1 holds. The converse, which is much harder, was first established by Wagner (1964). Later, Halin (1964) and Young (1971) gave shorter proofs (see also Ore, 1967, pp. 134–163). Thus, we have the following theorem.

Theorem 7-2 Conjectures C_1 and C_{33} are equivalent.

To understand this result, there is a related notion we should mention here. H is a *subcontraction* of G if H is (isomorphic to) a subgraph of a contraction G_1 of G. We denote this by $G > H$. The following theorem is due independently to Harary and Tutte (1965) and Wagner (1937).

Theorem 7-3 G is nonplanar if and only if $G > K_5$ or $G > K_{3,3}$.

One can show $G > K_n$ if and only if G can be contracted to K_n. Hence, we could restate Conjecture C_{33} as $\chi(G) \geq 5$ implies $G > K_5$. In proving that C_1 implies C_{33} one notes that, by the 4CC, G is nonplanar since $\chi(G) = 5$. Thus, by the preceding theorem, $G > K_5$ or $G > K_{3,3}$. Of course, the difficulty in proving that C_1 implies C_{33} involves showing that we actually have $G > K_5$.

We need to make some preliminary definitions in order to state a conjecture of Hajós (1961). We shall follow the approach of Berge (1973, pp. 350–352; see also Ore, 1967, p. 185).

Let $\mathcal{G}_q$ denote the class of all graphs G with $\chi(G) > q$. Consider the following three amalgamation operations:

(1) Add edges and vertices.
(2) Let G_1 and G_2 be two disjoint graphs with $[v_i, w_i] \in E(G_i)$, $i = 1, 2$. Form a new graph from the disjoint union of G_1 and G_2 by deleting both of the edges $[v_1, w_1]$ and $[v_2, w_2]$, adding an edge $[w_1, w_2]$, and identifying the two vertices v_1 and v_2 (see Fig. 7-1).
(3) Identify a pair of nonadjacent vertices to create a single vertex.

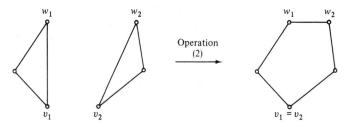

Figure 7-1

Lemma 7-1 If $G \in \mathcal{G}_q$ and any of the operations (1), (2), or (3) are performed on G, the resulting graph is still in $\mathcal{G}_q$.

PROOF Operations (1) and (3) are trivial. (Why?) Now suppose G is formed from G_1 and G_2 using operation (2) and assume that c is q-coloring of G. Let v be the vertex of G corresponding to v_1 and to v_2. If $c(v) \neq c(w_1)$, then c would induce a q-coloring of G_1. Hence, $c(v) = c(w_1)$. But $[w_1, w_2] \in E(G)$ so $c(w_1) \neq c(w_2)$. Therefore, $c(v) \neq c(w_2)$ so that c induces a q-coloring of G_2.

For example, then, any graph formed from copies of K_{q+1} using the above three operations must have chromatic number greater than q. The next result of Hajós (1961) is the converse statement.

Theorem 7-4 Any graph $G \in \mathcal{G}_q$ can be obtained from copies of K_{q+1} using operations (1), (2), and (3).

PROOF Suppose G is a simple graph of order n contradicting the theorem so $\chi(G) > q$ but G cannot be constructed from the K_{q+1} using operations (1), (2), and (3). Assume further that G has a maximal number of edges among all such graphs (with order n).

We claim that $\overline{G}$, the complement of G, is a disjoint union of complete graphs. Granting the claim for a moment, we go on to finish the proof. If $\overline{G} = K_{p_1} \cup \cdots \cup K_{p_k}$, where the union is vertex-disjoint, then $G = K_{p_1, \ldots, p_k}$ is a complete k-partite graph. Since $K_k \subset G$ but G is not constructable from K_{q+1} using (1), we must have $k \leq q$. Hence, $\chi(G) = k \leq q$, which is impossible.

Now let us verify the claim. Choose a component C of $\overline{G}$. If C is not complete, then there are three vertices u, v, w of C with $[u, v] \in E(\overline{G})$, $[v, w] \in E(\overline{G})$, $[u, w] \notin E(\overline{G})$. Now return to G. By maximality, $G \cup [u, v]$ and $G \cup [v, w]$, respectively, contain graphs G_1 and G_2 which are constructable from K_{q+1}. The subgraphs G_1 and G_2 necessarily contain the edges $[u, v]$ and $[v, w]$, respectively. Now take a pair of disjoint graphs G_1' and G_2' which are isomorphic to G_1 and G_2, respectively. Let $[u_1, v_1] \in E(G_1')$ correspond to $[u, v] \in E(G_1)$ and $[v_2, w_2] \in E(G_2')$ to $[v, w] \in E(G_2)$. Perform operation (2) on G_1' and G_2' with respect to these edges, identifying v_1 and v_2 and connecting u_1 and w_2 with an edge. Call the resulting graph G'. Now identify in G' any pair of vertices corresponding to the same vertex in G; this is just operation (3). The resulting graph G'' is a subgraph of G, and so G can be constructed using the three operations. This contradiction establishes the claim and, hence, the theorem.

Conjecture C_{34} Any graph obtained from K_5 using operations (1), (2), and (3) (called five-amalgamation) is nonplanar.

Theorem 7-5 Conjectures C_{34} and C_1 are equivalent.

PROOF By Hajós' theorem, G satisfies the hypotheses of C_{34} if and only if G is five-chromatic. But the statement "G is five-chromatic implies G is nonplanar" is plainly equivalent to Conjecture C_1.

7-3 CHAIN GROUPS AND THE 4CC

In this section, we shall present an exposition of some recent and very stimulating work of Tutte (1966; see also 1967b) which is in some sense a generalization of the Hadwiger Conjecture, C_{33} to finite projective geometries over $GF(2)$. We shall define projective geometries shortly, but first we need some preliminaries on the theory of "chain groups." This is related to the theory of matroids introduced by Whitney (1935) and greatly expanded by Tutte (1965) in his lectures on the subject.

In Sec. A-2 of the Appendix, we define a one-chain c in a graph G as a formal sum $\sum_{e \in E'} e$ for some subset E' of $E(G)$. The *support* of c, $\| c \|$, is the set E'. If H is any graph, we call H *eulerian* if every vertex in H has even degree. It is clear that $c \in C_1(G)$ is a one-cycle if and only if $G(\|c\|)$ is eulerian. The reader can easily check that H is eulerian if and only if it has a spanning circuit which covers every edge. Hence, algebraic cycles coincide with geometric circuits.

Two one-chains c and d in $C_1(G)$ are *orthogonal* if $\sum_{e \in E(G)} c(e)d(e) = 0$, where $c(e)$ denotes the coefficient (0 or 1) with which e occurs in c. Since $Z_1(G) \oplus Z_1^*(G) \cong C_1(G)$, it follows that $c \in C_1(G)$ is a cycle if and only if it is orthogonal to every coboundary and it is a coboundary if and only if it is orthogonal to every cycle.

Let us generalize the notions of zero-chains and one-chains. Suppose E is any finite set and let $GF(2)$ denote the finite field with two elements ($GF(2)$ is often denoted Z_2 since it is merely the integers Z modulo 2). We can form a vector space $C(E)$ over the field $GF(2)$ with basis E. An element $f \in C(E)$, called a *chain on E*, can be regarded as a function $f : E \to GF(2)$ which assigns a *coefficient* $f(e)$ equal to 0 or 1 to each element $e \in E$. If $E = \varnothing$, $C(E)$ is taken to be 0. If $f \in C(E)$, then f is uniquely determined by its coefficients $f(e)$, $e \in E$. Thus, f can be regarded as a formal sum $\Sigma f(e)$ (sum over all $e \in E$). If $f, g \in C(E)$, then $(f + g)(e) = f(e) + g(e)$. If $f \in C(E)$, define $\| f \| = \{ e \in E \mid f(e) = 1 \}$. Then $\| f + g \| = \| f \| \Delta \| g \|$, where "$\Delta$" denotes "symmetric difference" of sets, $S \Delta T = (S \cup T) - (S \cap T)$.

Let N be any subgroup of $C(E)$; thus, N is a subset of $C(E)$ and if $f, g \in N$, then $f + g \in N$. N is called a *chain group* on E. The elements of E are called the *cells* of N. If x is a cell of N, x is *filled* if $x \in \| f \|$ for some $f \in N$ and *empty* otherwise.

If G is a graph, $C_1(G) = C(E(G))$ and $C_0(G) = C(V(G))$. For an edge e and vertex v we define the *incidence number* $\eta(e, v) \in \mathrm{GF}(2)$ by

$$\eta(e, v) = \begin{cases} 0 & \text{if } v \text{ is not incident with } e \text{ or if } e \text{ is a loop} \\ 1 & \text{if } v \text{ is incident with } e \text{ and } e \text{ is not a loop} \end{cases}$$

We can give the boundary and coboundary operators in terms of incidence numbers. If $f \in C_1(G)$, then, for any $v \in V(G)$,

$$(\partial f)(v) = \sum_{e \in E(G)} \eta(e, v) f(e)$$

defines $\partial f \in C_0(G)$. Similarly, if $g \in C_0(G)$, then, for any $e \in E(G)$,

$$(\delta g)(e) = \sum_{x \in V(G)} \eta(e, v) g(v)$$

The following lemmas are easy consequences of the definitions and are left as exercises.

Lemma 7-2 An edge $e \in E(G)$ is a bridge if and only if there exists $f \in Z_1^*(G)$ such that $\| f \| = e$.

Lemma 7-3 An edge $e \in E(G)$ is a loop if and only if it is empty in $Z_1^*(G)$.

Let N be a chain group on E. A *coloring* of N is a pair $\{f, g\}$ of chains of N such that, for every cell $x \in E$, either $f(x) = 1$ or $g(x) = 1$. Call N *chromatic* if it has a coloring, and *achromatic* if it does not have a coloring. A chromatic chain group N is necessarily *full* (that is, N can have no empty cells). This algebraic notion of coloring will turn out to correspond both to four-colorings of vertices and three-colorings of edges.

Consider the following set of four vectors whose components are in $\mathrm{GF}(2)$: $a = (1, 1)$, $b = (1, 0)$, $c = (0, 1)$, and $d = (0, 0)$. Suppose G is a graph. An assignment α of the four "colors" a, b, c, and d to the vertices of G is then just an ordered pair (f_1, f_2) of zero-chains given by the formula $\alpha(v) = (f_1(v), f_2(v))$ for any $v \in V(G)$. The assignment α is a coloring of the vertices if and only if $\alpha(v) \neq \alpha(w)$ whenever v and w are adjacent. But $\alpha(v) \neq \alpha(w)$ if and only if either $f_1(v) \neq f_1(w)$ or $f_2(v) \neq f_2(w)$. Thus, if e is an edge joining v and w, either $(\delta f_1)(e) = 1$ or $(\delta f_2)(e) = 1$. This proves the following result.

Theorem 7-6 $\{f_1, f_2\}$ is a four-coloring of G if and only if $\{\delta f_1, \delta f_2\}$ is a coloring of $Z_1^*(G)$.

Corollary 7-1 Let G be a graph. Then G can be four-colored if and only if $Z_1^*(G)$ has a coloring.

Now we want to provide a similar equivalent for Tait-colorings. We shall use the three colors a, b, and c above.

Theorem 7-7 Let G be a cubic graph. Then G has a Tait-coloring if and only if $Z_1(G)$ has a coloring.

PROOF Suppose G has a Tait-coloring τ using a, b, and c. As before, we can represent the coloring by an ordered pair (f_1, f_2) of one-chains so that $\tau(e) = (f_1(e), f_2(e))$, $e \in E(G)$. Consider any vertex v and the three edges e_a, e_b, e_c incident with v and colored a, b, and c, respectively. Precisely, two of these edges have coefficient 1 in f_1 and precisely two have coefficient 1 in f_2 (this follows from the fact that all three edges have different colors and the color $d = (0, 0)$ is not used). Hence, f_1 and f_2 are one-cyles of G, and (again since d is not used) $\{f_1, f_2\}$ is a coloring of $Z_1(G)$.

Conversely, suppose $\{f_1, f_2\}$ is a coloring of $Z_1(G)$. If $e \in E(G)$, define $\tau(e) = (f_1(e), f_2(e))$. Then $\tau(e)$ is a, b, or c. Let $v \in V(G)$ and let e_1, e_2, and e_3 be its three incident edges, two of which may coincide if G has a loop at v. If two of these edges have the same color, the third would have to be colored d (to preserve the parity of the number of ones) since f_1 and f_2 are one-cycles in G. But we have not used color d so all three edges are colored differently. Hence, (f_1, f_2) is a Tait-coloring of G.

Let N be a chain group on E. The *rank* of N, $r(N)$, is the maximum number of linearly independent chains in N. (Recall that a set S of chains is *linearly independent* if no subset $T \subseteq S$ has sum zero, $\sum_{f \in T} f = 0$.) Of course, $r(N)$ is just the dimension of N as a vector space over GF(2). If N is a chain group on E, we define the *dual* chain group N^* on E by $N^* = \{f \in C(E) \mid f$ is orthogonal to every $g \in N\}$. It is easy to check that N^* *is* a chain group and that $C(E) \cong N \oplus N^*$, where "$\oplus$" denotes the direct sum of vector spaces. Put differently, every $g \in C(E)$ can be uniquely expressed as a sum $g = f + f^*$, where $f \in N$, $f^* \in N^*$, and "$+$" means sum in $C(E)$. The following results are trivial consequences of linear algebra:

$$r(N^*) = n - r(N), \quad \text{where} \quad n = |E| = r(C(E))$$

$$N^{**} = N$$

A bijection from one chain group to another is an *isomorphism* if it preserves sums. Since one can prove that $Z_1(G) \oplus Z_1^*(G) \cong C_1(G)$, it follows that $Z_1^*(G)$ is isomorphic to the chain group dual to $Z_1(G)$.

Let N be a chain group on E and $S \subseteq E$. If $f \in N$, define a chain f_S on S called the *restriction* of f to S by $f_S(x) = f(x)$ for any $x \in S$. Clearly, $\{f_S \mid f \in N\}$ is a chain group on S, and we denote it by $N \cdot S$ and call it the *reduction* of N to S. Since $\{f_S, g_S\}$ is a coloring of $N \cdot S$ whenever $\{f, g\}$ is a coloring of N, we have the following lemma.

Lemma 7-4 If $N \cdot S$ is achromatic, N is achromatic.

We can use N to define another chain group $N \times S$ on S, called the *contraction* of N to S, as follows:

$$N \times S = \{f_S \mid f \text{ is a chain on } N \text{ and } \| f \| \subseteq S\}$$

Lemma 7-5 Suppose $T \subseteq S \subseteq E$ and N is a chain group on E. Then $(N \cdot S) \cdot T = N \cdot T$ and $(N \times S) \times T = N \times T$.

PROOF Since $(f_S)_T = f_T$, the first equality holds. The second equality also uses the fact that $\| g \| \subseteq T$ implies $\| g \| \subseteq S$.

We need another technical result.

Theorem 7-8 Suppose $T \subseteq S \subseteq E$ and N is a chain group on E. Then

$$(a) \quad (N \times S) \cdot T = \{N \cdot [E - (S - T)]\} \times T$$

$$(b) \quad (N \cdot S) \times T = \{N \times [E - (S - T)]\} \cdot T$$

PROOF $(N \times S) \cdot T$ consists of the restrictions f_T of those chains f in N with $f(e) = 0$ for $e \in E - S$. Since $E - S = (E - (S - T)) - T$, $(N \times S) \cdot T = \{f_T \mid f \in N \text{ with } f(e) = 0 \text{ for } e \in (E - (S - T)) - T\} = (N \cdot (E - (S - T))) \times T$. Thus, (a) holds. The second equality follows from the first upon replacing S by $E - (S - T)$ since $E - [E - (S - T) - T] = E - (E - S) = S$.

A *minor* of a chain group N on E is any chain group of the form $(N \times S) \cdot T$, where $T \subseteq S \subseteq E$.

Theorem 7-9 Any reduction or contraction of N is a minor of N. A minor of a minor of N is a minor of N.

PROOF Since $N \cdot S = (N \times E) \cdot S$ and $N \times S = (N \times S) \cdot S$, the first statement follows. The second statement is an immediate consequence of Lemma 7-5 and Theorem 7-8.

Thus, being a minor is a reflexive and transitive relation. The notions of reduction and contraction correspond (intuitively, for now) to the ideas of subgraph and contraction. The following theorem furthers this belief.

Theorem 7-10 Let $S \subseteq E$ and N be a chain group on E. Then

$$(N \cdot S)^* = N^* \times S$$

$$(N \times S)^* = N^* \cdot S$$

PROOF Let f be any chain on S. There is a unique chain f' on E such that $\| f' \| \subseteq S$ and $f'_S = f$. Now f is orthogonal to every chain of $N \cdot S$ if and only if f' is orthogonal to every chain of N; that is, if and only if f is a chain of $N^* \times S$.

To establish the second equality, replace N by N^* in the first equality to obtain

$$(N^* \cdot S)^* = N^{**} \times S = N \times S$$

Now just take the dual of both sides.

Corollary 7-2 The minors of N^* are the duals of the minors of N.

Now let us apply these notions to graphs. If G is a graph and $S \subseteq E(G)$, let us write $G \cdot S$, instead of $G(S)$, for the smallest subgraph of G containing S. Thus, $E(G \cdot S) = S$ and $v \in V(G \cdot S)$ if and only if v is an endpoint of some e in S. We call $G \cdot S$ the *reduction* of G to S. If $G:S$ denotes the spanning subgraph of G with $E(G:S) = S$, then $G \cdot S$ is obtained from $G:S$ by deleting isolated vertices.

We construct another graph $G \times S$ called the *contraction* of G to S as follows: $E(G \times S) = S$. The vertices of $G \times S$ are in one-to-one correspondence with the components of $G:(E(G) - S)$. The endpoints of an edge e in $G \times S$ are the (not necessarily distinct) components of $G:(E(G) - S)$ containing the endpoints of e in G. Thus, $G \times S$ is obtained from G by contracting the components of $G:(E(G) - S)$ to single vertices, and we are justified in calling it the contraction of G to S.

The following results are easy consequences of the definitions (in fact, practically of the notation!). The last two identities can be derived from the first two using duality.

Theorem 7-11 $Z_1^*(G \cdot S) = Z_1^*(G) \cdot S$

$$Z_1^*(G \times S) = Z_1^*(G) \times S$$

$$Z_1(G \cdot S) = Z_1(G) \cdot S$$

$$Z_1(G \times S) = Z_1(G) \times S$$

An *irreducible* chain group is a full achromatic chain group which has no achromatic minor other than itself. A chain group N is *graphic* if $N = Z_1^*(G)$ for some graph G and *cographic* if $N = Z_1(H)$ for some graph H. A graphic (or cographic) chain group has graphic (or cographic) minors by Theorem 7-11.

We are about to rephrase Hadwiger's conjecture in terms of chain groups.

First, let us note that Hadwiger's conjecture can be equivalently restated: if G has no loops and cannot be four-colored, then some reduction of a contraction of G is K_5.

Conjecture C_{35} The only irreducible chain group which is graphic is $Z_1^*(K_5)$.

Our discussion has shown that this conjecture is equivalent to Conjecture C_{33}. There is a conjecture about Tait-colorings due to Tutte (1967b) which is very similar to Hadwiger's conjecture.

Conjecture T2 Let G be a cubic bridgeless graph with no Tait-coloring. Then G has the Petersen graph P as a subcontraction.

Clearly, Tutte's conjecture implies the 4CC. The reverse implication remains, to the best of our knowledge, an open question. Watkins (1967) has pointed out that the converse of Tutte's conjecture is false. Using Theorem 7-7 instead of Corollary 7-1, we can also equivalently phrase Tutte's conjecture in terms of chain groups.

Conjecture T2' The only irreducible chain group which is cographic is $Z_1(P)$.

7-4 PROJECTIVE GEOMETRIES AND THE 4CC

Now we want to relate chain groups to finite projective geometries. Let $PG(q, 2)$ denote the finite q-dimensional projective geometry over $GF(2)$ whose points may be identified with all vectors of the form $x_1, \ldots, x_{q+1}$, where each $x_i \in GF(2)$ and not all x_i are zero. Note that $PG(q, 2)$ consists of a $(q + 1)$-dimensional vector space over $GF(2)$ minus the origin. Thus, we can use vector space terminology when convenient. A k-space consists of a set Q of $k + 1$ linearly independent points together with all other points which are linear combinations of points in Q. Given any set B of points in $PG(q, 2)$, we define their *dimension* to be the dimension of the k-space they generate.

Let N be a chain group on a set E. An *embedding* of N in $PG(q, 2)$ is a function F mapping E onto a set of points in $PG(q, 2)$ satisfying the following condition. Let f be any nonzero chain on E. Then

$$\sum_{x \in E} (fx)(Fx) = 0$$

if and only if f is a chain of N^*. If N is cographic, say $N = Z_1(G)$, where G is a graph, then F is an embedding if and only if, for every subset $E_1 \subset E(G)$, FE_1 is a linearly dependent set of points precisely when E_1 contains a nonzero cycle. A point of FE_1 is repeated as many times as it is the image of a point in E_1. FE_1 is regarded as linearly dependent when it contains such points, and these points correspond to multiple edges in G. The embedding F is, in this case, called a *direct embedding* of G in $PG(q, 2)$. In an analogous way, we can define a *dual embedding* of G in $PG(q, 2)$ in terms of $Z^*_f(G)$ and a nonzero coboundary.

Let $k > 0$ be an integer. A set B of points in $PG(q, 2)$ is a *k-block* if its dimension is at least k and it includes at least one point from each subspace of $PG(q, 2)$ of dimension $q - k$. A k-block is *minimal* if no proper subset of it is a k-block. Let C be a nonempty subset of a k-block B. Any $(q - k)$-space which contains all points of C but no point of B which is independent of them is called a *tangent* of C. B is called a *tangential k-block* if every nonempty subset of B with dimension at most $q - k$ has a tangent in B. For example, any k-space is both a minimal and tangential k-block. It turns out that tangential implies minimal.

The following theorem is due to Tutte (1966), following an idea of Veblen (1913).

Theorem 7-12 Let F be an embedding in $PG(q, 2)$ of a chain group N on E. Then N is achromatic if and only if FE is a two-block.

We can go even further in this direction.

Theorem 7-13 Let F be an embedding in $PG(q, 2)$ of an achromatic chain group N on E. Then N is irreducible if and only if F is a one-to-one function from E onto FE and FE is a tangential two-block in $PG(q, 2)$.

In view of Conjecture C_{35}, we are led to investigate the tangential two-blocks in $PG(q, 2)$. Tutte (1966) has found all of them for $q \leq 5$.

Call a subset S of $PG(q, 2)$ *graphic* if it has the form $S = FE$ for F a direct embedding of some graph G and *cographic* if F is a dual embedding of some graph. Tutte found the following three tangential two-blocks:

1. The Fano block (the plane which has exactly seven points), which is neither graphic nor cographic;
2. The Desargues block (a three-dimensional two-block consisting of ten points lying in threes on ten lines in a Desargues configuration), which is graphic but not cographic and corresponds to K_5;
3. The Petersen block (this is the only five-dimensional, tangential two-block) which is cographic but not graphic and corresponds to the Petersen graph.
   In a private communication, Tutte has informed us that Datta has proved that there are no six-dimensional tangential two-blocks.

It is not known whether any higher dimensional tangential two-blocks exist. A complete classification of all tangential two-blocks would settle the 4CC. Suppose S is a two-block occurring as a direct embedding of some graph G with no four-coloring. We can modify S by a process of projection to obtain a tangential two-block S' which is the image of a direct embedding of a graph G' which is not four-colorable and is obtained from G by contracting some edges. The Hadwiger conjecture, if true, will force G' to be K_5, possibly with some multiple edges. Thus, a complete list of tangential two-blocks would enable us to check Hadwiger's conjecture and hence the 4CC.

Tutte is even willing to make the following assertion, which would imply the 4CC.

Conjecture T3 There are no other tangential two-blocks.

APPENDIX

A-1 INTRODUCTION

In order to preserve the flow of ideas we have sometimes avoided going into topological and graph-theoretic details. The appendix contains a number of these technical or tangential results. We hope that by presenting them here we shall convey some measure of the mathematical complexity underlying many areas of the book. We have also endeavored to collect a comprehensive list of books on graph theory.

The appendix contains six sections in addition to this one. Section A-2 deals with characterizations of planarity as well as some elementary algorithms. Section A-3 develops the theory of disconnecting sets in maximal planar graphs and leads to the elegant theorems of Steinitz and Whitney. Additional topics involving line graphs are treated briefly in the Sec. A-4. Section A-5 gives a few of the sets of sufficient conditions for guaranteeing the existence of a hamiltonian circuit, while Sec. A-6 presents a sketch of some results from Dirac's theory of critical graphs. Finally, in Sec. A-7, we discuss various topological generalizations of coloring problems.

A-2 PLANARITY

In this book we are concerned almost solely with planar graphs, but in the more general setting of graph theory it is often of great interest to determine whether or not a graph *is* planar. In this section, we present three well-known (noncombinatorial) necessary and sufficient conditions for planarity, formulated by Kuratowski (1930), MacLane (1937), and Whitney (1933c), respectively, and then we give an algorithm for determining whether or not certain graphs are planar.

Our first characterization of planarity is a sort of converse to Theorem 1-5.

Theorem A-1 (Kuratowski) Let G be a graph. Then G is nonplanar if and only if G contains a subgraph which is homeomorphic to either K_5 or $K_{3,3}$.

One half of the proof is clear. The other half—showing that if G is nonplanar, then G contains a subgraph homeomorphic to K_5 or $K_{3,3}$—is quite involved. The reader is referred to any standard text on graph theory (e.g., Busacker and Saaty, 1965, or Harary, 1969, pp. 109–112).

Our next criterion for planarity is due to MacLane. We will need to introduce several new concepts in order to explain his result. If G is embedded in the plane, then the boundaries of the regions produced are cycles, and these cycles have certain formal properties; e.g., every nonbridge edge belongs to exactly two of them. MacLane's characterization is a kind of converse to this observation. First, we need some machinery. For background, see Lefschetz or any basic text on algebraic topology.

Let G be a graph. A *one-chain* in G is a formal sum $c' = \sum_{e \in E'} e$ where E' is some subset of $E = E(G)$. If $c' = \sum_{e \in E'} e$ and $c'' = \sum_{e \in E''} e$, then $c' = c''$ if and only if $E' = E''$. We define the sum of c' and c'' by

$$c' + c'' = \sum_{e \in E' \Delta E''} e$$

where $E' \Delta E'' = E' \cup E'' - (E' \cap E'')$ is the *symmetric difference* of E' and E''. For example, $(e_1 + e_2 + e_4) + (e_1 + e_3 + e_4) = e_2 + e_3$. This addition is clearly commutative and associative. Moreover, $0 = \sum_{e \in \varnothing} e$ functions as an (additive) identity; that is, $0 + c' = c' = c' + 0$. Let $C_1 = C_1(G)$ denote the set of all one-chains. Since $c + c = 0$ for any $c \in C_1(G)$, C_1 has the structure of a mod 2 vector space. In fact, $C_1(G)$ is just the mod 2 vector space whose basis consists of the set $E(G)$. If G has no edges, $C_1(G) = 0$.

Similarly, define $C_0(G)$ to be the mod 2 vector space generated by the set $V(G)$ of vertices. Thus C_0 consists of formal sums of vertices (called *zero-chains*) under "mod 2" addition. We define a function $\partial: C_1 \to C_0$, called the *boundary operator*, by associating to an edge $e = [v, w]$ the sum $v + w$ (thus, if e is a loop, $\partial(e) = 0$). If $c = \sum_{e \in E'} e \in C_1(G)$, then

$$\partial(c) = \sum_{e \in E'} \partial(e)$$

This definition makes ∂ preserve addition: in other words, ∂ is a *homomorphism* of vector spaces over Z_2. We call $\partial(c)$ the *boundary* of c. We write $Z(G)$ for the kernel of ∂; that is, $Z(G)$ is the set of all one-chains whose boundary is 0. The elements of $Z(G)$ are called *one-cycles*. They correspond to circuits but are defined algebraically rather than geometrically. See Section 7.3 on eulerian graphs. Once again, $Z(G)$ is a mod 2 vector space since, if $\partial(z) = 0$ and $\partial(z') = 0$, then $\partial(z + z') = \partial(z) + \partial(z') = 0$.

Let V be any mod 2 vector space. A finite set $S \subset V$ is called *independent* if no element of S is a sum of other elements in S. A maximal independent set $B \subset V$ is called a *basis*. Maximal independent means that if $v \in V - B$, then $B \cup \{v\}$ is not independent. (A set is called *dependent* if it is not independent.) The vector spaces $C_1(G)$ and $C_0(G)$ have as bases $E(G)$ and $V(G)$, respectively, as we have already observed. A basis for $Z(G)$ is called a *cycle basis*. The number of elements in a cycle basis does not depend on the particular basis chosen; we call this number $\beta(G)$, the *cyclomatic* number of G. Every circuit, regarded as a set of edges, is a one-cycle. The next result determines $\beta(G)$, and its proof shows that $Z(G)$ has a basis of simple circuits.

Lemma A-1 Let G be a connected graph with n vertices and m edges. Then $\beta(G) = m - n + 1$.

First we need another lemma whose proof is left until later.

Lemma A-2 Every connected graph G contains a spanning subgraph H which is a tree.

H is called a *spanning tree* in G.

PROOF OF LEMMA A-1 Let H be a spanning tree in G and let G/H denote the graph obtained from G by contracting H to a point. Then G/H consists of a single vertex and as many loops as there are elements in $E(G) - E(H)$. Since H is a tree, it contains no cycles and hence $\beta(G) = \beta(G/H) = |E(G) - E(H)| = m - (n - 1) = m - n + 1$.

An addendum to the proof of this lemma, which shows that $Z(G)$ has a basis of circuits, can be obtained from the spanning tree H using a slightly different procedure. An edge e of G, which is not in H, is called a *chord* of H. Adding any chord e to the spanning tree H produces a graph $H \cup e$ with exactly one circuit, call it $Z(H, e)$ (see Fig. A-1), and this circuit must pass through the edge e. Now H has exactly $m - (n - 1) = m - n + 1$ chords and each one determines a circuit. Since the cycles determined by these circuits are all independent, it follows from the fact that $\beta(G) \leq m - n + 1$ that $Z(G)$ has a basis of circuits. Alternatively, we could simply note that any cycle Z is the mod 2 sum $Z = \Sigma Z(H, e)$, where the sum is taken over all edges e in Z which are chords in H.

Now we can state MacLane's planarity theorem (1937).

Theorem A-2 Let G be any connected nontrivial graph. Then G is planar if and only if (*) there is a cycle basis $Z_1, \ldots, Z_s$ for G and one additional cycle Z_0 such that $Z_0 = Z_1 + \cdots + Z_s$ and every nonbridge of G lies in Z_i for

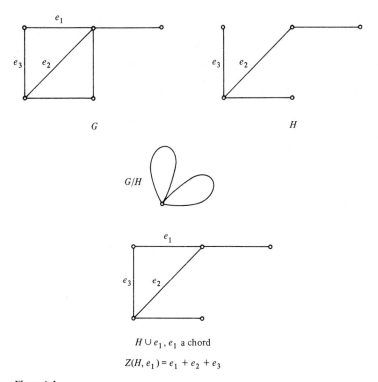

$H \cup e_1, e_1$ a chord

$Z(H, e_1) = e_1 + e_2 + e_3$

Figure A-1

precisely two of the indices i, $0 \le i \le s$. (Note that Z_0 may equal some Z_j, $1 \le j \le s$.)

PROOF One half is again easy. Suppose that G is planar, and take $Z_1, \ldots, Z_s$ as the boundaries of the finite regions of a map M obtained by drawing G in the plane and Z_0 as the boundary of the infinite face. Then $Z_1, \ldots, Z_s$ are independent and $Z_0 = Z_1 + \cdots + Z_s$. Since every nonbridge lies on the boundary of exactly two regions, we are done.

We sketch part of the converse. Suppose that (*) holds. One proves that G is planar by induction on the number m of edges. If $m \le 5$, G is certainly planar so we assume that the result holds for $m - 1$ and let G be any connected graph with $m \ge 6$ and satisfying (*). If G has no nonbridge edges, then G is a tree. Hence, we can suppose G has some nonbridge $e = [v, w]$.

Assume, without loss of generality, that e lies on Z_0 and Z_1. Let $G' = G - e$. Consider the one-cycles $Z'_1 = Z_0 + Z_1$, $Z'_j = Z_j$ $(2 \le j \le s)$. We claim that $Z'_2, \ldots, Z'_s$ forms an independent set of one-cycles in $Z(G')$ since $Z_2, \ldots, Z_s$ are independent in $Z(G)$. Moreover, since $\beta(G') = \beta(G) - 1 = s - 1$, the $Z'_2, \ldots, Z'_s$ form a basis of $Z(G')$. Also,

$$Z'_1 = Z_0 + Z_1 = (Z_1 + \cdots + Z_s) + Z_1 = Z_2 + \cdots + Z_s$$

Therefore, (*) holds for G' and hence, by induction, G' can be drawn in the plane.

It remains to show that G is planar. This amounts to proving it is possible to reintroduce the edge e so that it joins v and w, but does not produce any crossings. This can be done using the methods of algebraic topology.

Before giving Whitney's characterization of planarity, we need a bit more algebra. If v is a vertex in a graph G, define the *coboundary* $\delta(v)$ of v by

$$\delta(v) = \sum_{e \in E(v)} e$$

where $E(v) = \{e \in E(G) \mid e$ is incident once with $v\}$. Now extend linearly to any zero-chain

$$\delta(x) = \sum_{v \in V'} \delta(v) \quad \text{if} \quad x = \sum_{v \in V'} v$$

Note that if e is a loop at v, e does not belong to $E(v)$. We call $\delta(x)$ a *coboundary*, and the set of all coboundaries forms a Z_2 vector space which is contained in $C_1(G)$, for if $c^1 = \delta(x^1)$ and $c^2 = \delta(x^2)$, then $c^1 + c^2 = \delta(x^1 + x^2)$. This vector space is called the *cocycle space* and is denoted $Z^*(G)$.

Lemma A-3 Let G be a connected graph with n vertices. Choose any $n - 1$ vertices $v_1, \ldots, v_{n-1}$ in G. Then

$(**)$ $\begin{cases} (1) \ \delta(v_1), \ldots, \delta(v_{n-1}) \text{ forms a basis for } Z^*(G). \\ (2) \ \text{If } v_n \text{ is the remaining vertex, } \delta(v_n) = \delta(v_1) + \cdots + \delta(v_{n-1}). \\ (3) \ \text{If } e \text{ is any edge which is not a loop, then } e \text{ appears in precisely} \\ \quad\ \text{two of } \delta(v_1), \ldots, \delta(v_n). \end{cases}$

PROOF Statement (3) is trivial, and (1) follows from (2) since the coboundaries $\delta(v_1), \ldots, \delta(v_{n-1})$ are independent and since any coboundary $= \delta(x)$ is a sum of the form $\Sigma\delta(v)$. Therefore, it suffices to prove (2).

This is just a straightforward calculation:

$$\sum_{i=1}^{n-1} \delta(v_i) = \sum_{i=1}^{n-1} \sum_{e \in E(v_i)} e = \sum_{e \in E(v_n)} e = \delta(v_n)$$

The following elegant combinatorial formulation of planarity is due to Whitney (1933c). Let G and G^* be connected graphs. A *bijection* is a one-to-one onto function. Any bijection $W: E(G) \to E(G^*)$ induces an isomorphism $\overline{W}: C_1(G) \to C_1(G^*)$ by setting $\overline{W}(c) = \sum_{e \in E'} W(e)$ whenever $c = \sum_{e \in E'} e$. W is called a Whitney duality if $\overline{W}$ carries $Z(G)$ onto $Z^*(G^*)$. In this case, we often write e^* and c^* for $W(e)$ and $W(c)$. We also say that G^* is a Whitney dual of G.

Theorem A-3 Let G be a connected graph. Then G is planar if and only if G has a Whitney dual.

PROOF Suppose that G is planar. Let M be a map whose underlying graph is G and let $G^* = D(M)$. We claim that G^* is a Whitney dual of G with Whitney duality W given by $W(e) = \hat{e}$ where $\hat{e}$ is the orthogonal edge to e (see the definition of $D(M)$ in Sec. 1-2). By MacLane's theorem, $Z(G)$ has a basis $Z_1, \ldots, Z_s$, where Z_i is the boundary of a finite region R_i. But $Z_i^* = \delta(\hat{R}_i)$, $1 \leq i \leq s$, where $\hat{R}_i$ is the vertex of $G^* = D(M)$ corresponding to R_i. Therefore, $Z_1^*, \ldots, Z_s^*$ determines a basis of $Z^*(G^*)$ by Lemma A-3. Thus, W carries $Z(G)$ onto $Z^*(G^*)$ since it carries basis to basis.

Conversely, suppose that $W: E(G) \to E(G^*)$ is a Whitney duality. By Lemma A-3, G^* satisfies (**), and hence, using the isomorphism from $Z(G)$ onto $Z^*(G^*)$ provided by W, G satisfies MacLane's condition (*). Therefore, G is planar.

The three characterizations of planarity which we have given are nice theoretically but may be difficult to apply and slow to yield results. In general, given any finite problem (e.g., determining whether or not a particular graph G contains a subgraph homeomorphic to K_5 or $K_{3,3}$), we may solve the problem by simply considering all possibilities (e.g., listing every subgraph H of G and then checking whether H is homeomorphic to K_5 or $K_{3,3}$). However, this may be a very lengthy method, involving $n!$ or even more operations to solve a problem of "size" n. Numbers like these will grow too fast even for the largest computer. Therefore, it becomes not only preferable but essential to develop quicker and more efficient procedures for solving finite problems. Such a procedure is called an *algorithm*. The number of operations in an algorithmic solution to a problem of size n may be only 2^n or even a polynomial in n. (Algorithms of the latter sort are often called *good* algorithms because they grow more slowly as n is increased.)

Algorithms form an important aspect of graph theory. For example, there is a famous algorithm due to Ford and Fulkerson (1962) for determining the maximum flow in a network. Several algorithms are known for checking the planarity or nonplanarity of a given graph. Most of them are quite lengthy, but we shall describe one (due to Demoucron, Malgrange, and Pertuiset, 1964) that works for a restricted class of graphs.

Call a graph G *hamiltonian* if it has a circuit which passes exactly once through every vertex. Such a circuit is called a *hamiltonian circuit*. Draw some hamiltonian circuit in G as a polygon in the plane and represent all of the remaining edges in the graph as interior diagonals. Construct an auxilliary graph A according to the following scheme. The vertices of A are in one-to-one correspondence with the interior diagonals of the polygon and two vertices are adjacent in A if and only if the corresponding diagonals cross each other. Conclude that G is planar if A is bipartite and that G is not planar if A is not bipartite (see Fig. A-2).

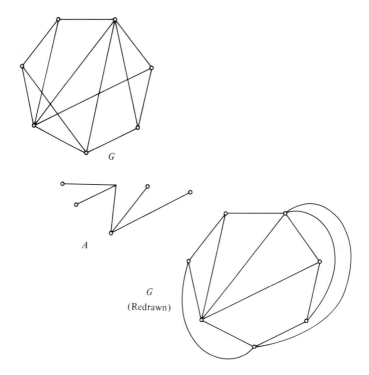

Figure A-2

Obviously, the validity of this algorithm requires proof.

Theorem A-4 Let G be hamiltonian and let A be an auxilliary graph constructed as above. Then G is planar if and only if A is bipartite.

PROOF Suppose that A is bipartite. Then we may partition the vertices of A into two classes so that no two vertices within the same class are adjacent. Represent G in the plane as follows. Draw the same hamiltonian circuit as before as a polygon and put one class of inner diagonals (corresponding to one class of vertices of A) inside the polygon and one class outside the polygon. Note that two diagonals cross on the outside if and only if they cross on the inside and that, of course, no inner diagonal can cross an outer diagonal.

Conversely, suppose that G is planar. Then draw G in the plane. The hamiltonian circuit which was used to construct A must be a simple closed curve, and every edge of G is either an inner or outer diagonal. The inner diagonals in the drawing provide one class of vertices in A and the outer diagonals provide the other class of vertices.

Now let us see how to algorithmically decide, for a hamiltonian graph G, whether or not G is planar. Since the problem of finding a specific hamiltonian circuit for G, even when we know that such a circuit exists, is highly nontrivial we shall further assume that we are given a specific hamiltonian circuit in G. This is equivalent to listing its vertices in the order in which they are encountered as we traverse the hamiltonian circuit.

Problem Given a graph G and hamiltonian circuit $v_1, \ldots, v_n$, determine whether G is planar.

Algorithm

Step 1 Choose n points on the circle $x^2 + y^2 = 1$ in clockwise order starting at the point $(0, 1)$ and call them $v_1, v_2, \ldots, v_n$, respectively.

Step 2 Every adjacent pair $[v_i, v_j]$ of vertices in G which are *not* joined by an edge in the hamiltonian circuit determine a vertex w_{ij} in the graph A.

Step 3 For each pair of vertices w_{ij} and w_{kl} in A, check to see if the corresponding line segments $[v_i, v_j]$ and $[v_k, v_l]$ cross. This means that $\{i, j\} \cap \{k, l\} = \varnothing$ and, moreover, that v_i, v_j lie on opposite sides of the line determined by $[v_k, v_l]$ and v_k, v_l on opposite sides of the line determined by v_i, v_j. (Such a linear algebra check can be rapidly performed by computer, with an effort or time which is approximately linear in n.)

Step 4 Join w_{ij} and w_{kl} with an edge in A if and only if the line segments $[v_i, v_j]$ and $[v_k, v_l]$ cross.

Step 5 Test whether A is bipartite or not by coloring some vertex red, then coloring all of its neighbors black, then all of their neighbors red, etc. Either A will be two-colored or an odd circuit will be found, in which case A is not bipartite. This test also requires constant times n steps.

Step 6 If A is two-colored by Step 5, the red vertices correspond to those edges of G (lying inside the circle $x^2 + y^2 = 1$) which should be "pulled outside" the circle in order to produce a planar embedding of G. (Again, see Fig. A-2.)

Although we have not done so here, it is clear that one can analyze such an algorithm and determine how long it takes to do the job for which it is designed.

When we stated Lemma A-2, we did not give a proof. There is a famous algorithm, due to Kruskal, called the "greedy algorithm," which shows that every connected graph G has a spanning tree H by actually constructing one. The idea is as follows. Choose any edge of G to start. Having chosen $k - 1$ edges of G so that no circuit is formed, choose a kth edge $e = [v, w]$ so that no circuit is determined by e. (This amounts to insuring that v and w do not belong to the same component of the subgraph we have already formed with the first $k - 1$ edges.) When no such edge e exists, we must have a spanning tree.

A-3 MAXIMAL PLANAR GRAPHS AND CONNECTIVITY

Let G be a graph. G is *k-connected* if G cannot be made either disconnected or trivial by the removal of any $k - 1$ vertices. The *connectivity* of G, $\kappa(G)$, is the largest k for which G is k-connected. G is *t-edge-connected* if it cannot be disconnected by the removal of any $t - 1$ edges. The edge-connectivity of G, $\kappa'(G)$, is the largest t for which G is *t-edge-connected* (see Fig. A-3). Whitney (1932a) first observed the following result; the proof is an exercise.

Lemma A-4 $\kappa(G) \leq \kappa'(G) \leq \delta(G)$.

Of course, an immediate consequence is that no planar graph has connectivity exceeding five.

The following lemma is essentially a restatement of Theorem 3-1.

Lemma A-5 The following are equivalent for a graph G with n vertices and m edges.
(1) G is maximal planar.
(2) $m = 3n - 6$.
(3) If G is drawn in the plane, then every region is triangular.

A simple characterization of k-connectedness is due to Whitney (1932a). If $v, w \in V(G)$, a v–w *path* is a path from v to w. Two v–w paths are called *disjoint* if they intersect only in v and w.

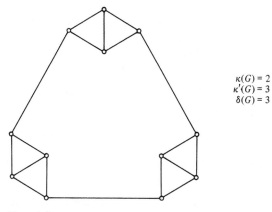

$$\kappa(G) = 2$$
$$\kappa'(G) = 3$$
$$\delta(G) = 3$$

Figure A-3

Lemma A-6 G is k-connected if and only if, for every pair of distinct vertices $v, w \in V(G)$, there are at least k disjoint v–w paths.

This result turns out to be nontrivial to prove. In fact, we need to invoke a basic and important theorem of graph theory which is due to Menger (1927). An elegant proof of this result has been provided by Dirac (1966).

The minimum number of vertices whose removal separates two nonadjacent vertices v and w is denoted $\kappa(v, w)$. The maximum number of disjoint v–w paths is written as $\mu(v, w)$. Note that $\kappa(v, w)$ is only defined when $[v, w] \notin E(G)$.

Theorem A-5 (Menger) $\kappa(v, w) = \mu(v, w)$.

Assuming Menger's theorem, we can prove Lemma A-6 easily. It is clear that $\kappa(G) = \min_{[v, w] \notin E(G)} \kappa(v, w)$. Now choose $v, w \in V(G)$ such that v and w are not adjacent. By Theorem A-5, $\kappa(v, w) = \mu(v, w)$ and, by assumption, $\mu(v, w) \geq k$. Therefore, $\kappa(G) \geq k$. If no such v and w exist, G is complete and the condition $\mu(v, w) \geq k$ for all v and w forces $\kappa(G) \geq k$.

On the other hand, if for some $v, w \in V(G)$ there are fewer than k disjoint v–w paths, then by Menger's theorem, v and w can be separated by fewer than k vertices, so $\kappa(G) < k$.

Let G be a connected graph and $S \subset V(G)$. Call S a *disconnecting* set if G–S is disconnected. We shall now obtain a very useful fact about disconnecting sets in maximal planar graphs.

Theorem A-6 Let G be any maximal planar graph drawn in the plane to produce a map M and let S be a disconnecting set. (We assume G has at least four vertices or else it has no disconnecting set.) Let G_1 and G_2 be two components of G–S. Then $G(S)$ contains a circuit Z which separates G_1 and G_2.

PROOF Choose $v_1 \in V(G_1)$ and $v_2 \in V(G_2)$. It suffices by the Jordan curve theorem to show that there is no simple arc a from v_1 to v_2 which lies entirely within $R^2 - G(S)$. But if there were such an arc a, then it is easy to see that a could be deformed to lie entirely within G–S, contradicting the fact that v_1 and v_2 belong to distinct components of G–S (see Fig. A-4).

In the above maximal planar graph G, $S = \{s_1, s_2, s_3, s_4\}$ is a separating set. Note that any simple arc a from v_1 to v_2 intersects $G(S)$ whose edges appear as heavy lines in Fig. A-4. On the other hand, the simple arc a' joining v_2 and v_7 misses $G(S)$ and can be deformed into a path $v_2, v_3, \ldots, v_7$ from v_2 to v_7 lying entirely within G–S.

Call S a *minimal disconnecting set* if G–S is disconnected but G–S' is connected for S', a proper subset of S.

Corollary A-1 Let S be a minimal disconnecting set for a maximal planar graph G_v. Then $G(S)$ is a cycle.

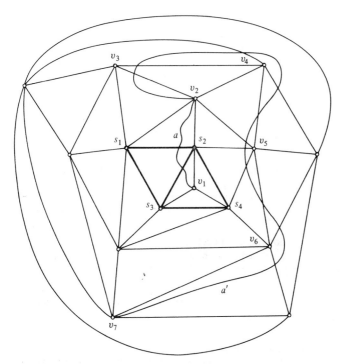

Figure A-4

Now we can prove a theorem due to Whitney (1932c).

Theorem A-7 Let G be a maximal planar graph with at least four vertices. Then G is three-connected.

PROOF Choose a disconnecting set S with precisely $\kappa(G)$ vertices. S is certainly minimal so, by Corollary A-1, $G(S)$ is a circuit. Therefore, $|S| \geq 3$.

Let us now digress momentarily to discuss graphs embedded in the sphere S^2. By adding a "point at infinity" the plane becomes the sphere and embedding of a graph in the plane becomes an embedding in the sphere. Conversely, if G is embedded in the sphere, then $S^2 - G$ consists of open connected regions and, in particular, it is nonempty. Removing any point x from $S^2 - G$ provides an embedding of G in $S^2 - x$, which is just the plane R^2. In fact, by choosing x in any specified region R of $S^2 - G$, we can embed G in the plane so that the boundary of R becomes the boundary of the infinite region.

In this way, a map M in the plane can be altered by first adding a point to put it into the sphere and then removing a different point to bring it back into the plane, producing a new map M' in which the boundary of the infinite region is the boundary of some finite region in M.

Obviously, M is not isomorphic to M' in the old sense, but we shall call them *spherically isomorphic* since they are isomorphic as maps in the sphere. Now we can give a very satisfying theorem due to Whitney (1933a).

Theorem A-8 Let G be a three-connected planar graph and let M and M' be any two maps produced by drawing G in the plane. Then M and M' are spherically isomorphic.

PROOF We already have the identity correspondence between vertices and edges of M and M'; we need only check that regions correspond to regions. Let R be any region in M and let Z be the boundary of R. Since G is three-connected, Z is a circuit. Let C' be the corresponding simple closed curve in M'.

Since G is planar and three-connected, $G–Z$ is connected and $G–C'$ is also connected. Hence, C' must be the boundary of a region R' in M', and R' is the region corresponding to R. Therefore, M and M' are spherically isomorphic.

Corollary A-2 Let M be a map. Then M is determined up to spherical isomorphism by $G = U(M)$ if G is three-connected.

There is an elegant result of Steinitz (see Steinitz and Rademacher, 1934; or Grünbaum 1967) which characterizes three-connected planar graphs and generalizes Theorem A-7.

Theorem A-9 G is planar and three-connected if and only if G is the one-skeleton of a convex three-dimensional polyhedron. (The one-skeleton of a polyhedron is just the graph determined by its vertices and edges.)

We should also mention here an interesting theorem due independently to Wagner (1936b), Fary (1948), and Stein (1951).

Theorem A-10 Let G be simple. Then G is planar if and only if G can be represented in the plane using straight-line segments. (That is, every planar graph can be *linearly* embedded in the plane.)

A-4 LINE GRAPHS REVISITED

The topic of line graphs, first raised in connection with the 4CC, has interesting topological properties and combinatorial generalizations, such as clique graphs.

Two results of Sedláček (1962) involve line graphs and topology.

Theorem A-11 A planar graph G has a planar line graph $L(G)$ if and only if no vertex in G has degree exceeding 4, and when a vertex has degree 4, then its removal must disconnect the graph.

Theorem A-12 If G is nonplanar, then $L(G)$ is nonplanar.

If G is a graph, we define a *clique* of G to be a maximal complete subgraph C of G, that is, C is complete but any strictly larger subgraph of G is not complete. The *clique graph* $K(G)$ of G has for vertices the cliques of G; two vertices are adjacent in $K(G)$ if and only if the corresponding cliques have a nonempty intersection in G. Call H a *clique graph* if $H = K(G)$ for some G. Roberts and Spencer (1971) have characterized clique graphs in terms of their complete subgraphs.

Theorem A-13 H is a clique graph if and only if there exists a family $\mathscr{F}$ of complete subgraphs of H such that

(1) $H = \bigcup_{F \in \mathscr{F}} F$.

(2) Whenever $G_i \cap G_j \neq \varnothing$ for all $G_i, G_j \in \mathscr{F}' \subseteq \mathscr{F}$, then $\bigcap_{F \in \mathscr{F}'} F \neq \varnothing$.

We can give a similar characterization of line graphs due to Krausz (1943; see also Harary, 1969, p. 74).

Theorem A-14 H is a line graph if and only if there exists a family $\mathscr{F}$ of complete subgraphs of H such that

(1) Every edge in H belongs to one and only one F in $\mathscr{F}$.
(2) No vertex in H belongs to more than two elements of $\mathscr{F}$.

Whitney (1932a) showed that if G has more than four vertices, then G is characterized by its line graph.

Theorem A-15 Let G_1 and G_2 be graphs and suppose that $L(G_1) \cong L(G_2)$ but $G_1 \not\cong G_2$. Then $L(G_1) \cong K_3$, $G_1 \cong K_{1,3}$, and $G_2 \cong K_3$ (or vice versa).

A-5 HAMILTONIAN GRAPHS REVISITED

In the text, we have not dealt with the existence of hamiltonian circuits for arbitrary graphs. In this section, we give a few of the most easily applied sufficient criteria for existence of a hamiltonian circuit in a graph. No useful characterization of such graphs is currently known.

Theorem A-16 (Dirac, 1952b) Let G be a simple graph. If G has n vertices and if, for every vertex v, deg $(v) \geq n/2$, where $n > 3$, then G is hamiltonian.

PROOF Consider a maximal (with respect to the number of edges) counter-example G to the theorem. By maximality, adding any edge to G must produce a hamiltonian graph. Thus, any two vertices of G which are not adjacent can be joined by a spanning chain (i.e., one which passes through every vertex) called a *hamiltonian chain*, in G. Since G is nonhamiltonian, it is not complete. Let v and w be two nonadjacent vertices, and let $v_1, \ldots, v_n$ be a hamiltonian chain joining them, where $v_1 = v$ and $v_n = w$. Since v and w are each adjacent to at least $n/2$ vertices, there must be some vertex v_k such that $[v, v_k] \in E(G)$ and $[w, v_{k-1}] \in E(G)$. But then

$$(v = v_1, \ldots, v_{k-1}, \quad w = v_n, v_{n-1}, \ldots, v_k, v)$$

is a hamiltonian circuit in G.

There is a similar result due to Ore (1960).

Theorem A-17 If G is a simple graph with n vertices, $n \geq 3$, and deg (v) + deg $(w) \geq n$ for all nonadjacent pairs v, w of vertices, then G is hamiltonian.

Although not every planar graph is hamiltonian (see Fig. A-5), Tutte (1956) has shown that four-connected planar graphs are hamiltonian. His result implies Whitney's, since, as we remarked, a triangle-free maximal planar graph with at least five vertices is four-connected.

Theorem A-18 Every four-connected planar graph is hamiltonian.

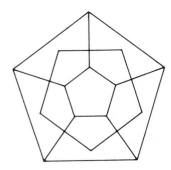

Figure A-5

A-6 CRITICAL GRAPHS

In Sec. 3-2, we mentioned minimal graphs which are planar graphs that are not four-colorable but such that any graph with fewer vertices is four-colorable. An interesting generalization of this notion is the concept of a critical graph. G is *k-critical* (or *k-vertex-critical*) if $\chi(G) = k$ but $\chi(G - v) < k$ for any vertex v in G (equivalently, $\chi(G - v) = k - 1$ for any $v \in V(G)$). Obviously, any k-critical graph G is connected. In fact, G has no *cutpoints* (vertices whose removal disconnects the graph).

Call a graph *nonseparable* if it is connected, nontrivial, and has no cutpoints.

Lemma A-7 Let G be k-critical. Then G is nonseparable.

PROOF It suffices to show that G has no cutpoints. So suppose $G - v$ is disconnected and let G_1 be a component of $G - v$ and G_2 the union of all other components. Let $\bar{G}_i$ be the full subgraph of G determined by G_i and $v (i = 1, 2)$. Clearly, $\bar{G}_1$ and $\bar{G}_2$ can be $(k - 1)$-colored and the two $(k - 1)$-colorings adjusted to produce a $(k - 1)$-coloring of G.

Lemma A-8 Let G be k-critical and suppose S is a separating set. Then $G(S)$ is *not* complete.

PROOF Suppose $G(S)$ is complete, say $G(S) = K_r$. Then $r \le k$ since $\chi(G) = k$, while if $r = k$ then $G(S)$ is a full subgraph of G with fewer vertices satisfying $\chi(G(S)) = \chi(G)$. This contradicts the fact that G is k-critical. Hence, $r < k$ so we can proceed as in the proof of Lemma A-7.

An easy result which is nevertheless worth stating is as follows.

Proposition A-1 Let G be a simple graph. If G is k-critical, then $\delta(G) \ge k - 1$.

PROOF If G has a vertex v of degree $\le k - 2$, then any $(k - 1)$-coloring of $G - v$ would extend to a $(k - 1)$-coloring of G.

Thus, the average degree $d(G)$ is at least $k - 1$ so $2m \ge n(k - 1)$, where m and n denote, respectively, the number of edges and vertices in G. We shall strengthen this result shortly.

A graph G is *edge-critical* if it is nontrivial and, for any edge e of G, $\chi(G - e) < \chi(G)$, or, equivalently, $\chi(G - e) = \chi(G) - 1$.

Lemma A-9 Let G be edge-critical. Then G is connected if and only if G has no isolated vertices. (A vertex is *isolated* if it is a connected component of G.)

PROOF Since G is edge-critical, it has at least one edge. Hence, if G is connected, it has no isolated vertices. Conversely, if G has no isolated vertices, then G

consists of components $G_1, \ldots, G_t$ each of which contains at least one edge. If $t \geq 2$, each G_i can be $(\chi(G) - 1)$-colored so G can be $(\chi(G) - 1)$-colored. Hence, $t = 1$ so G is connected.

The following lemma is left as an exercise.

Lemma A-10 Let G be a connected edge-critical graph. Then G is vertex-critical.

By Brooks' theorem a connected n-chromatic graph G with $\Delta(G) \leq n - 1$ is either complete or an odd cycle. Hence, G is edge-critical. The following proposition is a slight extension of this result.

Proposition A-2 Let G be a connected n-chromatic graph in which at most one vertex has degree exceeding $n - 1$. Then G is edge-critical.

PROOF Let e be any edge of G. It is easy to see that $\delta(G - e) \leq n - 2$ and that, in fact, sw $(G - e) \leq n - 2$. Hence, $\chi(G - e) \leq n - 1$ by Theorem 2-10 and so G is edge-critical.

We now give without proof an interesting result due to Dirac (1953).

Theorem A-19 Let G be a connected edge-critical graph with p vertices, q edges, and $\chi(G) = s$. Assume $s \geq 4$ and G is not complete. Then $2q \geq (s - 1)p + (s - 3)$.

Another interesting result of Dirac (1952b) describes circuits in critical graphs. The *circumference* of G is the length of the longest circuit of G.

Theorem A-20 Let G be k-critical with $k \geq 3$. Then G is hamiltonian or the circumference of G is at least $2k - 2$.

Let us close with a theorem that relates connectivity and criticality (see Ore, 1967, p. 1965).

Theorem A-21 Let G be k-critical, $k \geq 2$. Then G is $(k - 1)$-edge-connected.

PROOF Let G be k-critical. The cases $k = 2$ and $k = 3$ are trivial so we assume $k \geq 4$. Suppose G is not $(k - 1)$-edge-connected. Then there is a set $E' = \{e_1, \ldots, e_t\}$, $t \leq k - 2$ of edges in G and a nontrivial partitioning $V(G) = V_1 \cup V_2$ such that E' consists of all edges joining a vertex in V_1 to a vertex in V_2. Let $G_1 = G(V_1)$ and $G_2 = G(V_2)$.

Since G is k-critical, G_1 and G_2 can be $(k - 1)$-colored with colorings C_1 and C_2 each using the same $k - 1$ colors. Now simply alter the coloring C_1 of G_1 so that if $e = [v_1, v_2] \in E'$ with $v_i \in V_i$, then $C_1(v_1) \neq C_2(v_2)$.

A-7 TOPOLOGICAL AND OTHER GENERALIZATIONS

There are countless variations on the notion of chromatic number. In this final section, we survey a few of these generalizations and motivate an elegant conjecture which connects topology and graph coloring. Recall from Chap. 2 that sw(G) is the maximum of $\delta(H)$, H a subgraph of G. If G is a graph, call $W \subset V(G)$ k-degenerate if the induced subgraph $G(W)$ satisfies sw($G(W)$) $\leq k$. Thus, W is zero-degenerate if and only if it is an independent subset, and W is one-degenerate if and only if $G(W)$ is a forest. The *point-partition* number $\rho_k(G), k \geq 0$, is defined (by Lick and White, 1972) as the minimum number of subsets in a partition of $V(G)$ by k-degenerate subsets.

Clearly, $\rho_0(G) = \chi(G)$. The invariant $\rho_1(G)$ is also called the *point-arboricity* of G (see Chartrand, Kronk, and Wall, 1968). Many of the standard results for chromatic number generalize to ρ_k practically verbatim. For example, when $k = 0$, the next result is just the inequality $\chi(G) \leq |V(G)|$.

Lemma A-11 For any graph G, if $|V(G)| = p$ and $k \geq 0$, then $\rho_k(G) \leq \{p/(k + 1)\}$.

PROOF Partition $V(G)$ arbitrarily into r_i-element subsets W_i, $1 \leq i \leq t$, where $r_i \leq k + 1$ and all r_i are as close to $k + 1$ as possible. Then the W_i are all k-degenerate and $t \leq \{p/(k + 1)\}$.

The next theorem (due to Lick and White, 1970) is a straightforward extension of Lemma 6-1. For $k \geq 0$ let $\alpha_k(G)$ be the cardinality of the largest k-degenerate subset W of $V(G)$.

Theorem A-22 Let $k > 0$ and let G be any graph with n vertices. Then $n/\alpha_k(G) \leq \rho_k(G) \leq \{(\eta - \alpha_k(G))/(k + 1)\} + 1$.

PROOF Choose a partition $V(G) = W_1 \cup \cdots \cup W_r$ of $V(G)$ by k-degenerate subsets W_i with $r = \rho_k(G)$. Since $|W_i| \leq \alpha_k(G)$ for all i, the first inequality follows. To get the second result, take any k-degenerate subset S of $V(G)$ with $|S| = \alpha_k(G)$.

Since S is k-degenerate, $\rho_k(G - S) \geq \rho_k(G) - 1$. But by Lemma A-11, $\rho_k(G - S) \leq \{(\eta - \alpha_k(\varepsilon))/(k + 1)\}$ since $G - S$ has $n - \alpha_k(G)$ vertices. Thus, $\rho_k(G) \leq \rho_k(G - S) + 1 \leq \{(\eta - \alpha_k(G))/(k + 1)\} + 1$ as required.

The result of Szekeres and Wilf (1968) generalizes in a straightforward manner for ρ_k, as Lick and White showed.

Theorem A-23 For any G and any $k \geq 0$,

$$\rho_k(G) \leq 1 + \left\lceil \frac{\text{sw}(G)}{k + 1} \right\rceil$$

When $k = 1$, the theorem has a nice consequence since any planar graph has a Szekeres–Wilf number not exceeding 5.

Corollary A-3 If G is planar, then $\rho_1(G) \leq 3$.

Thus, the vertices of any planar graph can be partitioned into three disjoint subsets, each of which induces a forest. Theorem A-23 (for $k = 1$) and its corollary were originally proved by Chartrand and Kronk (1969).

Corollary A-3 can be regarded as the answer to a "packing" problem: "What is the maximum number of disjoint induced forests in a planar graph?" Other packing problems abound in graph theory. For example, the *coarseness* $\xi(G)$ of a graph G is defined to be the maximum number of edge-disjoint nonplanar subgraphs whose union is G.

Corresponding to any packing problem there are corresponding covering problems. For example, we may ask for the minimum number of planar subgraphs of G whose union is G; this is called the *thickness of* G (see Theorem 2-16). There is an even more direct analogue to point-arboricity: define the *arboricity* $a(G)$ of a graph G to be the minimum number of forests whose union is G. Note that we do not require the forests to be disjoint or induced, but we do insist that every edge, as well as every vertex, be covered.

The arboricity of any graph G is given by the following elegant result of Nash-Williams (1961).

Theorem A-24 For any graph G,

$$a(G) = \max_{2 \leq p \leq n} \left\{ \frac{m_p}{p-1} \right\}$$

where $n = |V(G)|$ and m_p denotes the maximum number of edges in any subgraph of G with p vertices.

PROOF One half of the theorem is easy to prove. For any graph H, $a(H) \geq \lceil |E(H)|/(|V(H)| - 1) \rceil$. Now $a(G) \geq \max_{H \subset G} a(H)$ and, for fixed $p = |V(H)|$, $a(H)$ is maximized by some H with a maximal number of edges. It now follows immediately that $a(G) \geq \max_{2 \leq p \leq n} \{m_p/(p-1)\}$, where n and m_p are defined as above.

While Nash-Williams' proof of equality does not provide a specific construction of the covering forests, Beineke (1964) has given such a decomposition for K_n and $K_{r,s}$, yielding directly that $a(K_n) = \{n/2\}$ and $a(K_{r,s}) = \{rs/(r + s - 1)\}$.

Applying Nash-Williams' result to a planar graph yields an interesting parallel to Corollary A-3.

Corollary A-4 If G is planar, then $a(G) \leq 3$.

PROOF By Theorem A-24, $a(G) = \max_{2 \le p \le n} \{m_p/(p-1)\}$. But any subgraph of G is planar, so $m_p \le 3p - 6 < 3(p-1)$. Hence, $a(G) \le 3$.

Returning to the topological point of view, we may ask: "How much can the condition of planarity be weakened without affecting the conclusion of the five-color theorem?" Specifically, let us define the *skewness* $\mu(G)$ of a graph G to be the minimum number of edges whose removal makes G planar. (If G is already planar, then $\mu(G) = 0$.)

We can now rephrase our question: "For how large an integer k does $\mu(G) \le k$ imply $\chi(G) \le 5$?" Note that we cannot hope to strengthen the four-color theorem in this way since $\mu(K_5) = 1$. But for five colors, planarity can be weakened somewhat, as the following theorem (Kainen, 1974a) shows.

Theorem A-25 If $\mu(G) \le 2$, then $\chi(G) \le 5$.

This result is the best possible since K_6 has skewness 3. Additional generalizations of coloring results can also be obtained by slightly relaxing embedding constraints using an analogous sort of skewness on higher-genus surfaces (see Kainen, 1974b, and also 1975).

In all known topological coloring problems the largest number of colors needed is exactly the largest number of vertices in a complete graph which satisfies the topological conditions. For example, the largest order of any complete graph which is planar is 4, and the four-color theorem shows that this is the largest number of colors needed to color any planar graph. Similar results hold for every topological surface, as we showed in Chap. 2 (see Theorem 2-21 and the discussion following Theorem 2-26).

We define a *topological class* to be an infinite set of graphs such that any graph which is a subgraph of or homeomorphic to a graph in the collection must itself be in the class. No counter-example is known to the following intriguing conjecture, which has also been called the metaconjecture.

Metaconjecture The largest number of colors needed to color any graph in a topological class is precisely the largest number of vertices in any complete graph belonging to the class.

BOOKS ON GRAPH THEORY

Alavi, Y., D. R. Lick, and A. T. White: "Graph Theory and Applications," Springer-Verlag, 1972.
Bari, R. A., and F. Harary (Eds.): "Graphs and Combinatorics Proceedings," Springer-Verlag, 1973.
Berge, C.: "Theory of Graphs and Its Applications," John Wiley, New York, 1961. (First appeared in French, 1958.)
——: "Principles of Combinatorics," Academic Press, 1971.
——: "Graphs and Hypergraphs," North-Holland, 1973.
—— and A. Ghouila-Houri: "Programming, Games and Transportation Networks," John Wiley, New York, 1965.
Bessonov, L. A.: "Elements of Graph Theory," Uchebn. Posobie Vses. Zaochn. Energ. Instituta, Moscow, 1964.
Biggs, N. L.: "Algebraic Graph Theory," Cambridge University Press, 1974.
Blažek, J. et al. (Eds.): Matematika (Geometric A. Teorie Grafi), Universita Kariova, Praha, 1970.
Bose, B. C., and T. A. Dowling (Eds.): "Combinatorial Mathematics and Their Applications," Univ. of North Carolina Press, Chapel Hill, 1969.
Busacker, R., and T. L. Saaty: "Finite Graphs and Networks," McGraw-Hill, 1965.
Cameron, P. J., and J. H. Van Lint: "Graph Theory, Coding Theory and Block Designs," Cambridge University Press, 1975.
Capobianco, M. et al. (Eds.): "Recent Trends in Graph Theory," Springer-Verlag, 1971.
Chartrand, G., and S. F. Kapoor (Eds.): "The Many Facets of Graph Theory," Springer-Verlag, 1969.
Chen, W. K.: "Applied Graph Theory," American Elsevier, 1972.
Christofides, Nicos: "Graph Theory; An Algorithmic Approach," Academic Press, 1975.
Deo, N.: "Graph Theory with Applications to Engineering and Computer Science," Prentice-Hall, 1974.
Dynkin, E. B., and W. A. Uspenski: "Multicolor Problems," D. C. Heath, Boston, 1952. (Original in German, English translation 1963.)
Erdös, P., and G. Katona (Eds.): "Theory of Graphs: Proceedings of the Colloquium held at Tihany, Hungary, in September, 1966," Academic Press, 1968.
Erdös, P., A. Renyi, and V. T. Sós (Eds.): "Combinatorial Theory and its Applications," North-Holland Publ. Co., Amsterdam, London, 1970.
Even, Shimon: "Algorithmic Combinatorics," Macmillan, 1973.
Fiedler, M. (Ed.): "Theory of Graphs and its Applications: Proceedings of the Symposium Smolenice, Czechoslovakia, June 17–20, 1963," Academic Press, New York, 1964.
Ford, L. R., and D. R. Fulkerson: "Flows in Networks," Princeton University Press, 1962.

Frank, H., and I. Frisch: "Communication, Transmission and Transportation Networks," Addison-Wesley, 1971.

Fulkerson, D. R. (Ed.): "Studies in Graph Theory, Parts I and II," The Mathematical Association of America, 1975.

Grossman, I., and W. Magnus: "Groups and Their Graphs," Random House, New York, 1964.

Grünbaum, B.: "Convex Polytopes," Interscience Publishers, New York, 1967.

Hall, M., Jr.: "Combinatorial Theory," Blaisdell Publications, 1967.

Hanani, H., N. Saver, and J. Schonheim (Eds.): "Combinatorial Structures and Their Applications," Gordon and Breach, New York, 1970.

Harary, F. (Ed.): "Graph Theory and Theoretical Physics," Academic Press, 1967.

———: "Graph Theory," Addison-Wesley, 1969.

———: "Proof Techniques in Graph Theory: Proceedings of the Second Ann Arbor Graph Theory Conference, February, 1968," Academic Press, New York, 1969.

———: "New Directions in the Theory of Graphs," Academic Press, 1973.

——— and L. Beineke (Eds.): "A Seminar on Graph Theory," Holt, Rinehart and Winston, New York, 1967.

——— R. Z. Norman, and D. Cartwright: "Structural Models: An Introduction to the Theory of Directed Graphs," John Wiley, New York, 1965.

——— and R. Z. Norman: "Graph Theory as a Mathematical Model in Social Science," University of Michigan, Ann Arbor, Michigan, 1953.

——— and E. M. Palmer: "Graphical Enumeration," Academic Press, 1973.

Harris, B.: "Graph Theory and its Application," Academic Press, 1975.

——— (Ed): "Graph Theory and Its Applications," Academic Press, 1976.

Henley, E. J., and R. A. Williams: "Graph Theory in Modern Engineering," Academic Press, 1973.

Hu, T. C.: "Integer Programming and Network Flows," Addison-Wesley, 1969.

Johnson, D. E., and J. R. Johnson: "Graph Theory with Engineering Applications," Ronald Press, 1972.

Kim, W. H., and R. T. Chien: "Topological Analysis and Synthesis of Communication Networks," Columbia University Press, New York, 1962.

König, D.: "Theorie Der Endlichen Und Unendlichen Graphen," Chelsea, New York, 1950.

Lefschetz, S.: *Application of Algebraic Topology* ("Graphs and Networks," "the Picard Lefschetz Theory," and "Feynman Integrals"), Springer-Verlag, 1975.

Lietzmann, W.: "Visual Topology," American Elsevier, New York, 1965.

Liu, C. L.: "Introduction to Combinatorial Mathematics," McGraw-Hill, 1968.

———: "Topics on Combinatorial Mathematics," Mathematical Association of America, 1972.

Macmahon, P. A.: "Combinatory Analysis," Chelsea, New York, 1960.

Malkevitch, J., and W. Meyer: "Graphs, Models and Finite Mathematics," Prentice-Hall, 1974.

Marshall, C. W.: "Applied Graph Theory," John Wiley, 1972.

Mayeda, W.: "Graph Theory," John Wiley, 1972.

Mirsky, L.: "Transversal Theory," Academic Press, 1971.

Moon, J. W.: "Topics on Tournaments," Holt, Rinehart and Winston, New York, 1968.

Nakanishi, N.: "Graph Theory and Feynman Integrals," Gordon, 1970.

Netto, E.: "Kombinatorik," Chelsea, 1927.

Ore, O.: "Theory of Graphs," Colloquium Publications, vol. 38, American Mathematical Society, Providence, Rhode Island, 1961.

———: "Graphs and Their Uses," Random House, New York, 1963.

———: "The Four Color Problem," Academic Press, 1967.

Percus, J. K.: "Combinatorial Methods," Courant Institute of Mathematical Sciences, New York University, 1969.

Proceedings of the Second Conference on Combinatorial Mathematics and its Applications, University of North Carolina Press, Chapel Hill, 1970.

Proceedings of the Second, Third, Fourth, Fifth, Sixth, and Seventh South East Conference on Combinatorics, Graph Theory, and Computing: Utilitus Mathematica Publishing, Winnipeg, Manitoba, Canada, 1970 through 1976.

Read, R. C. (Ed.): "Graph Theory and Computing," Academic Press, 1972.

Ringel, G.: "Farbungsprobleme auf Flachen und Graphen," VEB Deutsche Verlag der Wissenschaften, Berlin, 1959.

———: "Map Color Theorem," Springer-Verlag, 1974.

Riordan, J.: "Combinatorial Identities," John Wiley, 1968.

———: "Introduction to Combinatorial Analysis," John Wiley, 1968.

Rosenstiehl, P. (Ed.): "Theory of Graphs: International Symposium at Rome, July 1966," Gordon and Breach, New York, 1967.

Roy, B.: "Algèbre Moderne et Théorie des Graphes, Tome 1: Notions et résultats fondamentaux," Dunod, Paris, 1969.

———: "Algèbre Moderne et Théorie des Graphes, Tome 2: Applications et Problèmes Spécifiques," Dunod, Paris, 1969.

Ryser, H. J.: "Combinatorial Mathematics," John Wiley, 1963.

Saaty, T. L.: "Optimization in Integers and Related Extremal Problems," McGraw-Hill, 1970.

Sachs, H.: "Einfuhrung in die Theorie der Endlichen Graphen," Teubner, Leipzig, 1970.

———, H. J. Voss, and H. Walther (Eds.): "Beiträge zur Graphentheorie," Teubner, Leipzig, 1968.

Seshu, S., and M. B. Reed: "Linear Graphs and Electrical Networks," Addison-Wesley, Reading, Massachusetts, 1961.

Tutte, W. T.: "Connectivity in Graphs," University of Toronto Press, 1966.

———: "Recent Progress in Combinatorics: Proceedings of the Third Waterloo Conference on Combinatorics, May, 1968," Academic Press, New York, 1969.

Welsh, D. J. (Ed.): "Combinatorial Mathematics and its Applications: Proceedings at a Conference Held at the Mathematical Institute, Oxford, from 7–10 July, 1969," Academic Press, New York, 1969.

Welsh, D. J. and D. R. Woodall (Eds.): "Combinatories," Proceedings of the 1972 Oxford Combinatorial Conference, 1972.

White, A.: "Graphs, Groups and Surfaces," American Elsevier, 1973.

Wilson, R. J.: "Introduction to Graph Theory," Academic Press, New York and London, 1972.

Zykov, A. A.: *Theory of Finite Graphs* (Russian), Nauka, Novosibirsk, 1969.

BIBLIOGRAPHY

Aarts, J. M., and J. A. de Groot: A Case of Coloration in the Four Color Problem, *Nieuw Arch. Wis.*, vol. 11, pp. 10–18, 1963.

Alavi, Y., D. R. Lick, and A. T. White (Eds.): "Graph Theory and Applications," Springer Lecture Notes, no. 303, Springer-Verlag, 1972.

Alberston, Michael O.: "A lower bound for the independence number of a planar graph," *J. Comb. Th.*, vol. 13, 20, 84–93, 1976.

———: Finding an Independent Set in a Planar Graph, in *Graphs and Combinatorics* (R. Bari and F. Hanary, Eds.), Springer-Verlag, New York, 1974c, pp. 173–180.

——— and H. S. Wilf: Boundary Values in Chromatic Graph Theory, *Bull. Am. Math. Soc.*, 1974a, to appear.

——— and ———: Boundary Values in the Four Color Problem, *Trans. Am. Math. Soc.*, 1974b, submitted.

Allaire, F., and E. R. Swart: A Systematic Approach to the Determination of Reducible Configurations, *J. Comb. Theory*, ser. B., 1976, to appear.

Anderson, S.: "Graph Theory and Finite Combinatorics," Markham, Chicago, 1970.

Andrásfai, B., P. Erdös, and V. T. Sós: On the Connection between Chromatic Number, Maximal Clique and Minimal Degree of a Graph, *Discrete Math.*. vol. 8, pp. 205–218, 1974.

Appel, K. I.: Computing Configurations, *Notices Am. Math. Soc.*, vol. 20, abstract 704–A16, 1973.

——— and W. Haken: Every Planar Map is Four Colorable, *Bull. Am. Math. Soc.*, vol. 82, pp. 711–712, 1976a.

——— and ———: The Existence of Unavoidable Sets of Geographically Good Configurations, *Illinois J. Math.*, vol. 20, pp. 218–297, 1976b.

——— and ———, and J. Koch: "Every Planar Map is Four Colorable. Part I: Discharging," *Illinois Journal of Mathematics*, vol. 21, pp. 429–490, 1977.

——— and ———: "Every Planar Map is Four Colorable. Part II: Reducibility," *Illinois Journal of Mathematics*, vol. 21, pp. 491–567, 1977.

Auslander, L., T. A. Brown, and J. W. T. Youngs: The Imbedding of Graphs in Manifolds, *J. Math. Mech.*, vol. 12, pp. 629–634, 1963.

Ball, W. W. Rouse, and H. S. M. Coxeter: "Mathematical Recreations and Essays," Macmillan, New York, 1947.

Ballentine, J. P.: Problem E 1756, *Am. Math. Monthly*, vol. 73, p. 204, 1966.

Bari, R. A.: "Absolute Reducibility of Maps of at Most Nineteen Regions," Ph.D. Thesis, The Johns Hopkins University, 1966.

——: The Four Leading Coefficients of the Chromatic Polynomials $Q_n(u)$ and $R_n(x)$ and the Birkhoff–Lewis Conjecture, in "Recent Progress in Combinatorics" (W. T. Tutte, Ed.), Academic Press, New York, 1969.

——: Coefficients of u^{n-5} and u^{n-6} in the Q-chromial $Q_n(u)$, *Ann. N. Y. Acad. Sci.*, vol. 175, pp. 25–31, 1970.

——: Regular Major Maps of at Most 19 Regions and their Q-chromials, *J. Comb. Th.*, ser. B, vol. 12, pp. 132–142, 1972a.

——: Minimal Regular Major Maps with Proper 4 Rings, in "Graph Theory and Application," Springer Lecture Notes, Springer-Verlag, 1972b.

——: Chromatic Equivalence of Graphs, in "Graphs and Combinatorics," Springer Lecture Notes, Springer-Verlag, 1974.

Baltzer, R.: Eine Erinnerung an Möbius und seinen Freund Weiske, "Bericht über die Verhandlungen der Sachsischen Akademie der Wissenschaften zu Leipzig," *Math.-Nat. Kl.*, vol. 37, pp. 1–6, 1885.

Battle, J., F. Harary, Y. Kodama, and J. W. T. Youngs: Additivity of the Genus of a Graph, *Bull. Am. Math. Soc.*, vol. 68, pp. 565–568, 1962.

Baxter, Glen: A Necessary Condition for Four-Coloring a Planar Graph, *J. Comb. Th.*, vol. 1, pp. 375–384, 1966.

Bean, Dwight Richard: "Effective Coloration of Graphs," Ph.D. Thesis, University of California, San Diego, 1972, to appear.

Behzad, M. and G. Chartrand: "An Introduction to the Theory of Graphs," Allyn and Bacon, Boston, 1971.

Beineke, L. W.: Decompositions of Complete Graphs into Forests, *Magyar Tud. Akad. Mat Kutato Int. Kozl.*, vol. 9, pp. 589–594, 1964.

——: The Decomposition of Complete Graphs into Planar Subgraphs, in "Graph Theory and Theoretical Physics" (F. Harary, Ed.), Academic Press, London, 1967, pp. 139–154.

——: A Survey of Packings and Coverings of Graphs, in "The Many Facets of Graph Theory" (G. Chartrand and S. F. Kapoor, Eds.), Springer Lecture Notes, vol. 110, pp. 45–53, 1969.

—— and R. K. Guy: The Coarseness of $K_{m,n}$, *Canad. J. Math.*, 1974, to appear.

—— and F. Harary: The Genus of the n-cube, *Canad. J. Math.*, vol. 17, pp. 494–496, 1965a.

—— and ——: The Thickness of the Complete Graph, *Canad. J. Math.*, vol. 17, pp. 850–859, 1965b.

—— and ——, and J. W. Moon: On the Thickness of the Complete Bipartite Graph, *Proc. Cambridge Philos. Soc.*, vol. 60, pp. 1–5, 1964.

Beraha, S.: Ph.D. Thesis, Johns Hopkins University, 1974.

Berge, C.: Les Problemes de Coloration en Theorie des Graphes, *Publ. Inst. Stat. Univ. Paris*, vol. 9, pp. 123–160, 1960.

——: "Farbung von Graphen, deren Samtliche bzw. deren ungerade Kreise Starr Sind., Wissenschaftliche Zeitung," Martin Luther University, Halle Wittenberg, 1961, pp. 114–115.

——: "The Theory of Graphs and Its Application," Dunod, Paris, 1958 (in French); Methuen, London, 1962 (in English).

——: Sur Certains Hypergraphes Generalisant les Graphes Bipartites, in "Combinatorial Theory and its Applications, Balatonfured" (P. Erdos, A. Renyi, V. T. Sos, Eds.), North-Holland, Amsterdam and London, 1970, pp. 119–113.

——: "Graphs and Hypergraphs," American Elsevier, New York, 1973.

Berman, G., and W. T. Tutte: The Golden Root of a Chromatic Polynomial, *J. Comb. Th.*, vol. 6, pp. 301–302, 1969.

Bernhart, A.: Six-Rings in Minimal Five-Color Maps, *Am. J. Math.*, vol. 69, pp. 391–412, 1947.

——: Another Reducible Edge Configuration, *Am. J. Math.*, vol. 70, pp. 144–146, 1948.

Bernhart, F.: "Colored Plants, Recursive Sequences, and Map Coloring," Preprint, Kansas State University, 1972.

——: "A Three-Five Color Theorem," Preprint, University of Waterloo, 1973a.

————: An Approach to Combinatorial Planarity, *Notices AMS*, vol. 20, Abstract 701-05-17, 1973b.

————: "Topics in Graph Theory Related to the Five Color Conjecture," Doctoral Dissertation, Kansas State University, 1974a.

————: "On the Characterization of Reductions of Small Order," Research Report CORR 74-17, University of Waterloo, 1974b.

————: "Some Remarks on Open Sets of Colorings," Department of Combinatorics and Optimization, Research Report CORR 74-7, February 20, 1974c.

Biggs, N.: An Edge-Colouring Problem, *Am. Math. Monthly*, vol. 79, pp. 1018–1020, 1972.

————: "Algebraic Graph Theory," Cambridge Tracts in Mathematics, no. 67, Cambridge University Press, 1974.

Birkhoff, G. D.: A Determinant Formula for the Number of Ways of Colouring a Map, *Ann. Math.*, vol. 14, pp. 42–46, 1912.

————: The Reducibility of Maps, *Am. Journal Math.*, vol. 35, p. 115, 1913.

————: On the Number of Ways of Coloring a Map, *Proc. Edinburgh Math. Soc.*, vol. 2, pp. 83–91, 1930.

————: On the Polynomial Expressions for the Number of Ways of Coloring a Map, *Ann. Scuola Norm. Sup. Pisa*, vol. 2, pp. 85–103, 1934.

———— and D. Lewis: Chromatic Polynomials, *Trans. Am. Math. Soc.*, vol. 60, pp. 355–451, 1946.

Blanuša, , D.: Problem Cetiviju Boja, Hrvatsko Prirodoslovno Društvo Glasnik, *Mat. Fiz. Asty.*, ser. II, vol. 1, pp. 31–42, 1946. (In Croatian with French summary.)

Blatter, Christian: On the Algebra of the Four-Color Problem, *Enseignement Math.*, vol. 11, pp. 175–195, 1965.

Bondy, J. A.: "Balanced Colourings and the Four Colour Conjecture," *Proc. American Math. Soc.*, vol. 33 (1972), pp. 241–244, 1972.

Bose, B. C., and T. A. Dowling (Eds.): "Combinatorial Mathematics and Their Applications," University of North Carolina Press, Chapel Hill, 1969.

Brahana, H. R.: A Proof of Peterson's Theorem, *Ann. Math.*, ser. 2, 19, p. 59, 1917.

————: The Four Color Problem, *Am. Math. Monthly*, vol. 30, pp. 234–243, 1923.

————: Regular Maps on an Anchor Ring, *Am. J. Math.*, vol. 48, pp. 225–240, 1926.

Brooks, R. L.: On Colouring the Nodes of a Network, *Proc. Cambridge Philos. Soc.*, vol. 37, pp. 194–197, 1941.

Brown, W. G., and H. A. Jung: On Odd Circuits in Chromatic Graphs, *Acta. Math. Acad. Sci. Hung.*, vol. 20, pp. 129–134, 1969.

———— and J. W. Moon: Sur les Ensembles de Sommets Independants dans les Graphes Chromatiques Minimaux, *Canad. J. Math.*, vol. 21, pp. 274–278, 1969.

Brylawski, T.: Reconstructing Combinatorial Geometries, in "Graphs and Combinatorics," Springer Lecture Notes, Springer-Verlag, 1974.

Busacker, R., and T. L. Saaty: "Finite Graphs and Networks: An Introduction with Applications," McGraw-Hill, New York, 1965.

Cain, Bryan E.: "A Two-Color Theorem for Analytic Maps in R^n," Preprint, Iowa State University, 1973.

Cairns, E. S.: "Introductory Topology," Ronald Press, New York, 1961.

Capobianco, M., J. B. Frechen, and M. Krolik (Eds.), "Recent Trends in Graph Theory," Springer-Verlag, Berlin, 1971.

Carlitz, L.: The Number of Colored Graphs, *Canad. J. Math.*, vol 15, pp. 304–312, 1963.

Cartwright, D., and F. Harary: The Number of Lines in a Digraph of each Connectedness Category, *SIAM Review*, vol. 3, pp. 309–314, 1961.

———— and ————: On Colorings of Signed Graphs, *Elem. Math.*, vol. 23, pp. 85–89, 1968.

Cayley, A.: On the Colouring of Maps, *Proceedings of the London Mathematical Society*, vol. 9, p. 148, 1878. See also *Proceedings of Royal Geographical Society*, vol. 1, p. 259, 1879.

Chartrand, G., and D. Geller: Uniquely Colorable Planar Graphs, *J. Comb. Th.*, vol. 6, pp. 271–278, 1969.

———— and ————, and S. Hedetniemi: A Generalization of the Chromatic Number, *Proc. Cambridge Phil. Soc.*, vol. 64, pp. 265–271, 1968.

————, ————, and ————: Graphs with Forbidden Subgraphs, *J. Comb. Th.*, ser. B, vol. 10, pp. 12–41, 1971.

———— and S. F. Kapoor (Eds.): "The Many Facets of Graph Theory," Springer-Verlag, Berlin, 1969.

———— and H. V. Kronk: The Point-Arboricity of Planar Graphs, *J. London Math. Soc.*, vol. 44, pp. 612–616, 1969.

————, ————, and C. E. Wall: The Point-Arboricity of a Graph, *Israel J. Math.*, vol. 6, pp. 168–175, 1968.

Choinacki, C. A.: A Contribution to the Four Color Problem, *Am. J. Math.*, vol. 64, pp. 36–54, 1942.

Chuard, J.: Le Problème des Quatre Couleurs, *L'Ens. Math.*, vol. 22, pp. 373–374, 1923a.

————: Quelques Propriétes des Reseaux Cubiques Traces sur Une Sphère, *Comptes Rendus, Acad. Sci. Paris*, vol. 176, pp. 73–75, 1923b.

————: Les Reseaux Cubiques et le Problème des quarte Couleurs, *Mem. Soc. Vaudoise Sci. Nat.*, vol. 25, no. 4, pp. 41–101, 1932.

Chvátal, V.: A Note on Coefficients on Chromatic Polynomials, *J. Comb. Th.*, vol. 9, pp. 95–96, 1970a.

————: The Smallest Triangle-Free 4-Chromatic 4-Regular Graph, *J. Comb. Th.*, vol. 9, pp. 93–94, 1970b.

Cook, S. A.: The Complexity of Theorem-Proving Procedures, in "Proceedings of the Third ACM Symposium on Theory of Computing," 1971, pp. 151–158.

Cooper, E. D.: Combinatorial Map Theory, *J. Comb. Th.*, ser. B, 1974, to appear.

Coxeter, H. S. M.: Map-Coloring Problems, *Scripta Math.*, vol. 23, pp. 11–25, 1958.

————: The Four-Color Map Problem, *Math. Teacher*, vol. 52, pp. 283–289, 1959.

Dantzig, George B.: "Linear Programming and Extensions," Princeton University Press, Princeton, New Jersey, 1963. "Datta," Ph.D. Thesis, Ohio State University.

Debruijn, N.: A Color Theorem for Infinite Graphs and a Problem in the Theory of Relations, *Nederl. Akad. Wetensch. Proc. Ser. A, 54 Indag. Math.*, vol. 13, p. 371, 1951.

Democron, G., Y. Malgrange, and R. Pertuiset: Graphes Planaires, *Revue Francaise de Rech. Operat.*, vol. 30, pp. 33–47, 1964.

Descartes, B.: A Three Color Problem, *Eureka*, April, 1947a.

————: Network-Colorings, *Math. Gazette*, vol. 32, pp. 67–69, 1947b.

————: Solution, *Eureka*, March, 1948.

————: Solution to Problem 4526, *Am. Math. Monthly*, vol. 61, p. 352, 1954.

———— and R. Descartes: La Coloration des Cartes, *Eureka*, vol. 31, pp. 29–31, 1968.

Dewdney, A. K.: The Chromatic Number of a Class of Pseudo-2-Manifolds, *Manuscripta Math.*, vol., 6, pp. 311–320, 1972.

De Werra, D.: On Some Combinatorial Problems Arising in Scheduling, *Canad. Op. Res. Soc.*, ser. J, vol. 8, pp. 165–175, 1970.

Dirac, G. A.: Note on the Colouring of Graphs, *Math. Zeitschr.*, vol. 54, pp. 347–353, 1951.

————: A Theorem of R. L. Brooks and a conjecture of H. Hadwiger, *Proc. London Math. Soc.*, ser. 3, vol. 7, pp. 161–195, 1957a.

————: Some Theorems on Abstract Graphs, *Proc. London Math. Soc.*, ser. 3, vol. 2, pp. 69–81, 1952b.

————: The Structure of k-Chromatic Graphs, *Fund. Math.*, vol. 40, pp. 42–55, 1953.

————: Theorems Related to the Four Colour Conjecture, *J. London Math. Soc.*, vol. 29, pp. 143–149, 1954.

————: Circuits in Critical Graphs, *Monatsh. Math.*, vol. 59, pp. 178–187, 1955.

————: Map Colour Theorems Related to the Heawood Colour Formula, *J. London Math. Soc.*, vol. 31, pp. 460–471, 1956.

————: A Property of 4-Chromatic Graphs and Some Remarks on Critical Graphs, *J. London Math. Soc.*, vol. 27, pp. 85–92, 1952a.

————: Short Proof of a Map-Colour Theorem, *Canad. J. Math.*, vol. 9, pp. 225–226, 1957b.

————: Trennende Knotenpunktmengen und Reduzibilitat Abstrackter Graphen mit Anwendung auf das Vierfarbenproblem, *J. fur Math.*, vol. 204, pp. 116–131, 1960.

————: On the Four-Colour Conjecture, *Proc. London Math. Soc.*, ser. 3, vol. 13, pp. 193–218, 1963a.

————: Percy John Heawood, *J. London Math. Soc.*, vol. 38, pp. 263–277, 1963b.

————: On the structure of 5- and 6-Chromatic Abstract Graphs, *J. fur Math.*, vol. 214, pp. 43–52, 1964a.

————: Valence-Variety and Chromatic Number, *Wiss. Z. Martin Luther Univ. Halle-Wittenberg*, vol. 13, pp. 59–63, 1964b.

————: Short Proof of Menger's Graph Theorem, *Mathematika*, vol. 13, pp. 42–44, 1966.

————: On the Structure of k-Chromatic Graphs, *Proc. Cambridge Phil. Soc.*, vol. 63, pp. 683–691, 1967.

———— and S. Schuster: A Theorem of Kuratowski, *Indag. Math.*, vol. 6, pp. 243–248, 1954.

Dynkin, E. B., and W. A. Uspenski: "Multicolor Problems," D. C. Heath, Boston, 1963.

Edmonds, J.: A Combinatorial Representation for Polyhedral Surfaces, *Notices Amer. Math. Soc.*, vol. 7, p. 646, 1960.

Erdös, P.: Problems and Results in Chromatic Graph Theory, in "Proof Techniques in Graph Theory," (F. Harary, Ed.), Academic Press, 1969, pp. 27–35.

———— and A. Hajnal: On a Property of Families of Sets, *Acta Math. Acad. Sci. Hungary*, vol. 12, pp. 87–123, 1961.

———— and ————: On Chromatic Number of Graphs and Set-Systems, *Acta Math. Acad. Sci. Hungary*, vol. 17, pp. 61–99, 1966.

———— and G. Katona (Eds.): "Theory of Graphs," Akad. Kiado, Budapest, 1968.

———— and R. Rado, A Construction of Graphs Without Triangles Bearing Preassigned Order and Chromatic number, *J. London Math. Soc.*, vol. 35, pp. 445–448, 1960.

————, Renyi, and Turán-Sus (Eds.): "Combinatorial Theory and Its Applications, Balantonfured," North-Holland, Amsterdam and London, 1970.

Errera, A.: Une Contribution au Probleme des Quatre Couleurs, *Bull. de la Soc. Math. de France*, vol. 53, p. 42, 1925.

Ershov, A. P., and G. I. Kozhukhin: Estimates of the Chromatic Number of Connected Graphs, *Dokl. Akad. Nauk*, vol. 142, pp. 270–273, 1962; also *Trans. Soviet Math.*, vol. 3, pp. 50–53, 1962.

Fary, I.: On Straight Line Representation of Planar Graphs, *Acta Sci. Math. Szeged*, vol. 11, pp. 229–233, 1948.

Faulkner, G. B., and D. H. Younger: Non-Hamiltonian Cubic Planar Maps, *Discrete Math.*, vol. 7, pp. 67–74, 1974.

Fiamcik, J., and E. Jucovic: Colouring the Edges of a Multigraph, in "Combinatorial Theory and Its Applications, Balatonfured" (P. Erdös, Renyi and Turán-Sus, Eds.), North-Holland, Amsterdam, 1970, pp. 601–606.

Fiedler, M. (Ed.): "Theory of Graphs and its Applications: "Proceedings of the Symposium at Smolenice, Czechoslovakia, June 17–20, 1963," Academic Press, New York, 1964.

Finck, H. J.: Uber die Chromatischen Zahlen eines Graphen und Seines Komplements, I and II, *Wiss. Z. T. H. Ilmenau*, vol. 12, pp. 243–251, 1966.

————: On the Chromatic Number of a Graph and its Complement, in "Theory of Graphs: Proceedings of the Colloquium held at Tihany, Hungary, in September, 1966." (P. Erdös and G. Katona, Eds.), Academic Press, 1968, pp. 99–113.

———— and H. Sachs: Uber eine von H. S. Wilf Angegebene Schranke fur die Chromatische zahl Endlicher Graphen, *Math. Nachr.*, vol. 39, pp. 373–386, 1969.

Fisk, S.: Combinatorial Structure on Triangulations I: The Structure of Four Colorings, *Adv. in Math.*, vol. 11, pp. 326–338, 1973a.

————: Combinatorial Structures on Triangulations II: Local Colorings, *Adv. in Math.*, vol. 11, pp. 339–350, 1973b.

————: Some Topological Approaches to the Four Color Problem, *Bull. Am. Math. Soc.*, vol. 79, no. 5, p. 1051, 1973c.

————: Combinatorial Structures on Triangulations III: Coloring with Regular Polyhedra, *Adv. in Math.*, vol. 12, pp. 296–305, 1974.

Folkman, J., and D. R. Fulkerson: Edge Colorings in Bipartite Graphs, in "Combinatorial Mathematics and Their Applications" (B. C. Bose and T. A. Dowling, Eds.), University of North Carolina Press, Chapel Hill, 1969.

Fournier, J. C. and M. Las Vergnas: Une Classe d'Hypergraphes Bichromatiques, *Discrete Math.*, vol. 2, pp. 407–410, 1972.

———— and ————: Une Classe d'Hypergraphes Bichromatiques. II, *Discrete Math.*, vol. 7, pp. 99–106, 1974.

Franklin, P.: The Four Color Problem, *Am. J. Math.*, vol. 44, pp. 225–236, 1922.

————: A Six Colour Problem, *J. Math. Phys.*, vol. 13, pp. 363–369, 1934.

————: Note on the Four Color Theorem, *J. Math. Phys.*, vol. 16, pp. 172–184, 1938.

————: The Four Color Problem, *Scripta Math.*, vol. 6, pp. 149–156 and 197–210, 1939.

Fréchet, M., and K. Fan: "Initiation to Combinatorial Topology," Prindle, Weber and Schmidt, Boston, Massachusetts, 1967.

Frink, O.: A Proof of Petersen's Theorem, *Ann. Math.*, vol. 27, pp. 491–495, 1926.

Gallai, T.: Kritische Graphen, 1 and 11, *Publ. Math. Inst. Hungarian Acad. Sci.*, ser. A, vol. 8, pp. 165–192, 1963; vol 9, pp. 373–395, 1964.

————: On Directed Paths and Circuits, in "Theory of Graphs" (P. Erdös and G. Katona, Eds.), Akademiai Kiado, Budapest, 1968, and Academic Press, New York, 1968.

Gerencser, L.: On Coloring Problems (Hungarian), *Mat. Lapok.*, vol. 16, pp. 274–277, 1965.

Goldbeck, B. T.: "8-Rings in Minimal Maps," Ph.D. Thesis, University of Oklahoma, 1957.

Golovina, L. I., and I. M. Yaglom: "Induction in Geometry," D. C. Heath, Boston, 1963.

Greenwell, D.: Semi-uniquely n-Colorable Graphs, in "Proceedings of Second Louisiana Conference in Combinatorics and Graph Theory," (Mullin, Reid, Roselle, and Thomas, Eds.), University of Manitoba Press, 1971.

Greenwood, R. E., and A. M. Gleason: Combinatorial Relations and Chromatic Graphs, *Canad. J. Math.*, vol. 7, pp. 1–7, 1955.

Gross, Jonathan L., and Seth R. Alpert: Branched Coverings of Graph Imbeddings, *Bull. Am. Math. Soc.*, vol. 79, p. 5, September, 1973a.

———— and ————: "The Topological Theory of Current Graphs," IBM Report No. RC 4252, March 2, 1973b.

———— and Thomas W. Tucker, "Quotients of Complete Graphs: Revisiting the Heawood Map-Coloring Problem," 1974, to appear.

Grötzsch, H.: Ein Dreifarbensatz fur Dreikreisfreie Netze auf der Kugel, *Wiss. Z. Martin-Luther Univ. Halle-Wittenberg Math. Naturwiss. Reihe*, vol. 8, pp. 109–120, 1958.

Growney, W.: "Edge Conjugation and Coloration in Cubic Maps," Ph.D. Thesis, University of Oklahoma, 1970.

Grünbaum, B.: Grötzsch's Theorem on 3-Colorings, *Michigan Math. J.*, vol. 10, pp. 303–310, 1963.

————: "Convex Polytopes," Interscience Publishers (Division of John Wiley), 1967.

————: A Problem in Graph Colorings, *Am. Math. Monthly*, vol. 77, pp. 1088–1092, 1970a.

————: Higher-Dimensional Analogs of the Four-Color Problem and Some Inequalities for Simplicial Complexes, *J. Comb. Th.*, vol. 8, pp. 147–153, 1970b.

————: Acyclic Colorings of Planar Graphs, *Israel J. Math.*, vol. 14, pp. 390–408, 1973.

Gupta, R. P.: The Chromatic Index and the Degree of a Graph, *Notices Am. Math. Soc.*, vol. 13, no. 6, 1966.

————: "Bounds on the Chromatic and Achromatic Numbers of Complementary Graphs," Mimeo. series, no. 577, University of North Carolina, Chapel Hill, 1968.

————: Independence and Covering Numbers of Line Graphs and Total Graphs, in "Proof Techniques in Graph Theory" (F. Harary, Ed.), Academic Press, New York, 1969.

Gustin, W.: Orientable Imbedding of Cayley Graphs, *Bull. Amer. Math. Soc.*, vol. 69, pp. 272–275, 1963.

Guthrie, F.: Note on the Colouring of Maps, *Proc. Roy. Soc. Edinburgh*, vol. 10, p. 729, 1880.

Guy, R. K., and L. W. Beineke: The Coarseness of the Complete Graph, *Canad. J. Math.*, vol. 20, pp. 888–894, 1968.

————, H. Hanani, N. Sauer, and J. Schonheim (Eds.): "Combinatorial Structures and Their Applications," Gordon and Breach, New York, 1970.

Hadwiger, H.: Uber eine Klassifikation der Streckenkomplexe, *Vierteljschr. Naturforsch. Ges., Zurich*, vol. 88, pp. 133–142, 1943.

————: Ungelöste Probleme, *Element Math.*, vol. 12, pp. 61–62, 1957.

————: Ungelöste Probleme, *Element Math.*, vol. 13, pp. 127–128, 1958.

Hajos, G.: Über eine Konstruktion nicht *n*-Farbbarer Graphen, *Wiss. Z. Martin Luther Univ. Halle-Wittenberg Math. Naturwiss. Reihe*, vol. 10, pp. 116–117, 1961.

Haken, W.: "On Shimamoto's Construction," Notes for 1971 Seminar at University of Illinois, 1971.

————: An Existence Theorem for Planar Maps, *J. Comb. Th.*, ser. B, vol. 13, pp. 180–184, 1973a.

————: Geographically Good Configurations, *Notices, Am. Math. Soc.*, vol. 20, Abstract 704-A17, 1973b.

Halin, R.: Bemerkungen über Ebene Graphen, *Math. Ann.*, vol. 153, pp. 38–46, 1964a.

————: On a Theorem of Wagner Related to the Four-Color Problem (German), *Math. Ann.*, vol. 153, pp. 47–62, 1964b.

————: A Colour Problem for Infinite Graphs, in "Combinatorial Structures and Their Applications," Gordon and Breach, New York, 1970, p. 123.

————: Coloring Seven Circuits, in "Graphs and Combinatorics," Springer-Verlag, 1974.

Hall, D. W.: On Golden Identities for Constrained Chromials, *J. Comb. Th.*, vol. 11, pp. 287–298, 1971.

————: Coloring Seven-Circuits in, "Graphs and Combinatorics," vol. 706, pp. 273–290, Springer-Verlag, 1977.

————: Chromatic Polynomials and Graph Coloring, *N. Y. State Math. Teachers J.*, vol. 23, 1973.

————and D. Lewis: Coloring Six Rings, *Trans. Am. Math. Soc.*, vol. 64, pp. 184–191, 1948.

————, J. W. Siry, and B. R. Vanderslice: The Chromatic Polynomial of the Truncated Icosahedron, *Proc. Am. Math. Soc.*, vol. 16, pp. 620–628, 1965.

Hall, P.: On Representations of Subsets, *J. London Math. Soc.*, vol. 10, pp. 26–30, 1935.

Hammer, F. D.: The Four-Color Theorem for Touching Pennies, *Am. Math. Monthly*, vol. 73, pp. 485–486, 1976.

Harary, F.: A Complementary Problem on Non-Planar Graphs, *Math. Mag.*, vol. 35, pp. 301–304, 1962.

———— (Ed.): "Graph Theory and Theoretical Physics," Academic Press, London, 1967.

————: "Graph Theory," Addison-Wesley, 1969a.

————: The Four Color Conjecture and Other Graphical Diseases, in "Proof Techniques in Graph Theory," (F. Harary, Ed.), Academic Press, 1969b.

————: On the Enumeration Program for Trying to Settle the FCC, *Notices Am. Math. Soc.*, vol. 20, Abstract 704-A12, 1973.

————and S. Hedetniemi: The Achromatic Number of a Graph, *J. Comb. Th.*, 1974, to appear.

————, ————, and R. W. Robinson: Uniquely Colorable Graphs, *J. Comb. Th.*, vol. 6, 1969.

————and W. T. Tutte: A Dual Form of Kuratowski's Theorem, *Canad. Math. Bull.*, vol. 8, pp. 17–20, 373, 1965.

Heawood, P. J.: Map-Colour Theorems, *Quart. J. Math.*, Oxford ser., vol. 24, pp. 332–338, 1890.

————: On the Four-Colour Map Theorem, *Quart. J. Math.*, vol. 29, pp. 270–285, 1898.

————: On Extended Congruences Connected with the Four-Colour Map Theorem, *Proc. London Math. Soc.*, vol. 33, pp. 252–286, 1932.

————: Failures in Congruences Connected with the Four-Colour Map Theorem, *Proc. London Math. Soc.*, vol. 40, pp. 189–202, 1936.

————: Note on a Correction in a Paper on Map-Congruences, *J. London Math. Soc.*, vol. 18, pp. 160–167, 1943; vol. 19, pp. 18–22, 1944.

Hedetniemi, S.: Disconnected-Colorings of Graphs, in "Combinatorial Structures and Their Applications," Gordon and Breach, New York, 1970, pp. 163–167.

Heesch, H.: "Untersuchungen zum Vierfarbenproblem," Bibliog. Institut, AG, Mannheim, 1969.

————: Chromatic Reduction of the Triangulations T_e, $e = e_5 + e_7$, *J. Comb. Th.*, ser. B, vol. 13, pp. 46–55, 1972.

Heffter, L.: Uber das Problem du Nachbargebiete, *Ann. Math.*, vol. 38, pp. 477–508, 1891.

Hell, P.: Absolute Planar Retracts and the Four Color Conjecture, *Notices Am. Math. Soc.*, vol. 20, Abstract 73T-A142, 1973.

Herzog, M. and J. Schonheim: The B_r Property and Chromatic Numbers of Generalized Graphs, *J. Comb. Th.*, ser. B, vol. 12, pp. 41–49, 1972.

Heuchenne, C.: Sur le Critère de Chromaticité de Hajos–Ore, *Bull. Soc. Roy. Liège*, vols. 1–2, pp. 10–13, 1968.

Hobbs, A. M.: A Survey of Thickness, in "Recent Progress in Combinatorics" (W. T. Tutte, Ed.), Academic Press, New York, 1969, pp. 255–264.

Hursch, J. L.: Edge-3-Colorability of Certain Graphs, *Notices Am. Math. Soc.*, vol. 22, 1955 (copy of manuscript 1975.)

Isaacs, R.: Infinite Families of Nontrivial Trivalent Graphs which are not Tait Colorable, *Am. Math. Monthly*, vol. 82, pp. 221–239, 1975.

Iyer, M. R., and V. V. Menon: On Coloring the $n \times n$ Chessboard, *Am. Math. Monthly*, vol. 73, pp. 721–725, 1966.

Izbicki, I. H.: Verallgemeinerte Farbenzahlen, in "Beitrage zur Graphentheorie" (H. Sachs, H. Voss, H. Walther, Eds.), Teubner, Leipzig, 1968, pp. 81–84.

Jacques, A.: "Constellations et Proprietes Algebriques des Graphes Topologiques," Ph.D. Thesis, University of Paris, 1969.

Johnson, E. L.: "A Proof of the Four-Coloring of the Edges of a Regular Three-degree graph, O.R.C. 63–28 (R.R.)," Mimeo. report, Operations Research Center, University of California, 1963.

———: A Proof of 4-Coloring the Edges of a Cubic Graph, *Am. Math. Monthly*, 1966.

Johnson, Lee W.: Upper Bounds for Vertex Degrees of Planar 5-Chromatic Graphs, *Trans. Am. Math. Soc.*, vol. 181, pp. 53–59, 1973.

Kainen, P. C.: On the Chromatic Number of Certain 2-Complexes, in "Proceedings of the Third Southeastern Conference on Combinatorics, Graph Theory and Computing, Florida Atlantic University, Boca Raton, 1972," University of Manitoba Press, 1972a.

———: "On the Chromatic Number of a Pinched Manifold," Preprint, 1972b.

———: Relative Colorings of Graphs, *J. Comb. Th.*, ser. B, vol. 14, pp. 259–262, 1973a.

———: Thickness and Coarseness of Graphs, *Abh. Math. Sem. Univ. Hamburg*, vol. 39, pp. 88–95, 1973b.

———: A Generalization of the 5-Color Theorem, *Proc. Am. Math. Soc.*, vol. 45, pp. 450–453, 1974a.

———: An Introduction to Topological Graph Theory, in "Graphs and Combinatorics," Lecture Notes in Mathematics, Springer-Verlag, New York, 1974b.

———: Chromatic Number and Skewness, *J. Comb. Th.*, ser. B, vol. 18, pp. 32–34, 1975.

Karp, R. M.: Reducibility among Combinatorial Problems, in "Complexity of Computer Computations" (R. E. Miller and J. W. Thatcher, Eds.), Plenum Press, New York, 1972, pp. 85–104.

———: The Probabilistic Analysis of Some Combinatorial Search Algorithms, in "Algorithms and Complexity New Directions and Recent Results" (J. F. Traub, Ed.), Academic Press, New York, 1976, pp. 1–19.

Kelly, J. B., and L. M. Kelly: Paths and Circuits in Critical Graphs Coloring, *Am. J. Math.*, vol. 76, pp. 786–792, 1954.

Kelly, P.: "Computer Assisted Study of Planar Maps," Ph.D. Thesis, University of Waterloo, 1974.

Kempe, A. B.: How to Color a Map with Four Colours without Coloring Adjacent Districts the Same Color, *Nature*, vol. 20, p. 275 1879; *Nature*, vol. 21, pp. 399–400, 1880.

———: On the Geographical Problem of the Four-Colors, *Am. J. Math.*, vol. 2, pp. 193–200, 1879.

König, D.: "Theorie der Endlichen und Unendlichen Graphen," Leipzig, 1936. (Reprinted Chelsea, New York, 1950.)

Kotzig, A.: Pair Hajos Graphs (Slovakian), *Casopis Pest. Mat.*, vol. 88, pp. 236–241, 1963.

———: Colouring of Trivalent Polyhedra, *Canad. J. Math.*, vol. 17, pp. 659–664, 1965.

———: Transformations of Edge-Colourings of Cubic Graphs, *Discrete Math.*, vol. 11, pp. 391–399, 1975a.

———: "Change Graphs of Edge-Colourings of Planar Cubic Graphs," University of Montreal, 1975b.

Kramer, F., and H. Kramer: Une Probleme de Coloration des Sommets d'un Graphe, *C. R. Acad. Sci. Paris*, vol. 268, pp. A 46–48, 1969.

Krausz, J.: Démonstration Nouvelle d'une Théorème de Whitney sur les Réseaux, *Mat. Fiz. Lapok*, vol. 50, pp. 75–89, 1943.

Kronk, H. V.: An Analogue to the Heawood Map-Coloring Problem, *J. London Math. Soc.*, ser. 2, vol. 1, pp. 550–552, 1969.

————: The Chromatic Number of Triangle-Free Graphs, in "Graph Theory and Applications" (Y. Alavi, D. R. Lick, and H. T. White, Eds.), Springer Lecture Notes, no. 303, Springer-Verlag, 1972, pp. 179–181.

———— and John Mitchem: On Dirac's Generalization of Brooks' Theorem, *Canad. J. Math.*, vol. XXIV, no. 5, pp. 805–807, 1972a.

———— and ————: The Entire Chromatic Number of a Normal Graph is at Most Seven, *Bull. Am. Math. Soc.*, vol. 78, no. 5, September, 1972b.

———— and A. T. White: A 4-Color Theorem for Toroidal Graphs, *Proc. Am. Math. Soc.*, 1974, to appear.

Kuratowski, K.: Sur le Problème des Courbes Gauches en Topologie, *Fund. Math.*, vol. 15, pp. 271–283, 1930.

Lee, L.: "Chromatically Equivalent Graphs," Ph.D. Dissertation, George Washington University, 1974.

Lefschetz, S.: *Application of Algebraic Topology* ("Graphs and Networks," "the Picard Lefschetz Theory," and "Feynman Integrals"), Springer-Verlag, 1975.

Lempel, A., S. Even, and I. Cederbaum: An Algorithm for Planarity Testing of Graphs, in "Theorie des Graphes, Rome I.C.C." (P. Rosenthiehl, Ed.), Dunod, Paris, 1967, pp. 215–232.

Levow, R. B.: Realizable Sets of Border Colorings, *Notices Am. Math. Soc.*, Abstract 709–A 21, November, 1973.

————: Relative Colorings and the Four-Color Conjecture, *Am. Math. Monthly*, vol. 81, pp. 491–492, 1974.

Lick, Don R., and Arthur T. White: k-Degenerate Graphs, *Canad. J. Math.*, vol. XXII, no. 5, pp. 1082–1096, 1970.

Liu, C. L.: "Introduction to Combinatorial Mathematics," McGraw-Hill, 1968.

Lovasz, L.: Graphs and Set Systems, in "Beiträge zur Graphentheorie" (H. Sachs, H. J. Voss, and H. Walther, Eds.), Teubner, 1968a, pp. 99–106.

————: On Chromatic Number of Finite Set Systems, *Acta Math. Acad. Sci. Hungary*, vol. 19, pp. 59–67, 1968b.

————: Normal Hypergraphs and the Perfect Graph Conjecture, *Discrete Math.*, vol. 2, pp. 253–267, 1972.

————: Independent Sets in Critical Chromatic Graphs, *Studia Sci. Math. Hungary*, vol. 8, pp. 165–168, 1973.

————: "Three Short Proofs in Graph Theory," Preprint, 1974.

Lucas, E.: The Geographical Four Colour Problem, *Revue Scientifique*, vol. 3, no. 6, pp. 12–17, 1883.

————: "Récréations Mathématiques," A. Blanchard, Paris, 1892.

MacLane, S.: A Structural Characterization of Planar Combinatorial Graphs, *Duke Math. J.*, vol. 3, pp. 340–472, 1937.

McWorter, W. A.: Coloring a Generalized Map, Prob. E. 1989, *Am. Math. Monthly*, pp. 788–789, September, 1968. (University of British Columbia E. 1989, p. 589, 1967.)

Maghout, K.: Sur la Détermination des Nombres de Stabilité et du Nombre Chromatique d'un Graphe, *C. R. Acad. Sci., Paris*, vol. 248, pp. 3522–3523, 1959.

Malek, M., and Z. Skupién: On the Maximal Planar Graphs and the Four Colour Problem, *Prace Mat.*, vol. 12, pp. 205–209, 1969.

Marathe, C. R.: On the Dual of a Trivalent Map, *Am. Math. Monthly*, vol. 68, May, 1961.

Marble, G., and Matular: "Computational Aspects of 4-Coloring Planar Graphs," Research Report, University of Wisconsin, 1972.

May, K. O.: The Origin of the Four-Color Conjecture, *Isis*, vol. 56, pp. 346–348, 1965.

Mayer, J.: Le Problème des Régions Voisines sur les Surfaces Closes Orientables, *J. Comb. Th.*, vol. 6, pp. 177–195, 1969.

Mayer, J. J.: *Combinatorial Theory*, ser. B, vol. 16 (1974).

Meek, B. L.: Some Results on k-Maps, *Math. Gazette*, vol. LII, February–December, 1968.

Melnikov, L. S., and V. G. Vizing: New Proof of Brooks' Theorem, *J. Comb. Th.*, vol. 7, pp. 289–290, 1969.

Menger, K.: Zur Allgemeinen Kurventheorie, *Fund. Math.*, vol. 10, pp. 96–115, 1927.

Meyer, W.: Five Coloring Planar Maps, *J. Comb. Th.*, ser. B., vol. 13, pp. 72–82, 1972.

Minty, G. J.: A Theorem on *n*-Coloring the Points of a Linear Graph, *Am. Math. Monthly*, vol. 69, pp. 623–624, 1962.

————: A Theorem on Three-Coloring the Edges of a Trivalent Graph, *J. Comb. Th.*, vol. 2, pp. 164–167, 1967.

Mitchem, John: On the Point-Arboricity of a Graph and its Complement, *Canad. J. Math.*, vol. 23, pp. 287–292, 1971.

————: Every Planar Graph has an Acyclic 8-Coloring, *Duke Math. J.*, vol. 41, 1974.

Moon, J. W.: Disjoint Triangles in Chromatic Graphs, *Math. Mag.*, November–December, 1966.

Moore, E. F.: Symmetries of 3-Regular 3-Connected Planar Graphs, *Notices Am. Math. Soc.*, vol. 20, Abstract 704-65, 1973.

Motzkin, T. S.: Colorings, Cocolorings and Determinant Terms, in "Theory of Graphs, Rome I.C.C." (P. Rosenstiehl, Ed.), Dunod, Paris, 1967, pp. 253–254.

Mycielski, J.: Sur le Coloriage des Graphs, *Colloq. Math.*, vol. 3, pp. 161–162, 1955.

Nash-Williams, C. St. J. A.: Edge-Disjoint Spanning Trees of Finite Graphs, *J. London Math. Soc.*, vol. 36, pp. 445–450, 1961.

Nesetril, J.: On *k*-chromatic graphs (Russian), *Comm. Math. Univ. Carol.*, vol. 7, pp. 3–10, 1966.

Nordhaus, E., and J. Gaddum: On Complementary Graphs Coloring, *Am. Math. Monthly*, vol. 63, pp. 175–177, 1956.

Ore, O.: Note on Hamilton Circuits, *Am. Math. Monthly*, vol. 67, p. 55, 1960.

————: "The Theory of Graphs," A.M.S. Colloquium Publications, 1962.

————: "The Four Color Problem," Academic Press, New York, 1967.

———— and G. J. Stemple: Numerical Methods in the Four Color Problem, in "Recent Progress in Combinatorics," (W. T. Tutte, Ed.), Academic Press, New York, 1969.

———— and ————: Numerical Calculations on the Four-Color Problem, *J. Comb. Th.*, vol. 8, pp. 65–78, 1970.

Pannwitz, E.: Review, *Jahrbuch über die Fortschritte der Math.*, vol. 58, p. 1204, 1932.

Peck, J. E. L., and M. R. Williams: Examination Scheduling, Algorithm 286, *Communications of the Association for Computing Machinery*, vol. 9(6), 1966.

Petersen, J.: Die Theorie der Regulären Graphen, *Acta Math.*, *Stockholm*, vol. 15, pp. 193–220, 1891. See also *Intermed. Math.*, vol. 5, pp. 225–227, 1898; *ibid*, vol. 6, pp. 36–38, 1899.

Pósa, L.: A Theorem Concerning Hamiltonian Lines, *Publ. Math. Inst. Hung. Acad. Sci.*, vol. 7, pp. 225–226, 1962.

Rabin, M. O.: Probabilistic Algorithm, in "Algorithms and Complexity New Directions and Recent Results," (J. F. Traub, Ed.), Academic Press, New York, 1976.

Ratib, I., and C. E. Winn: Generalization of a Reduction of Errera in the Four Color Problem, *Inter. Cong. Math. Oslo*, vol. 2, pp. 131–139, 1936.

Read, R. C. (Ed.): The Number of *k*-Coloured Graphs, Canad. J. Math., vol. 12, pp. 410–414, 1960.

————: An Introduction to Chromatic Polynomials, *J. Comb. Th.*, vol. 4, pp. 52–71, 1968.

————: "Graph Theory and Computing," Academic Press, 1972.

Rector, R. W.: "Fundamental Linear Relations for the Seven-ring," Ph.D. Thesis, 1973.

Reynolds, C. N.: On the Problem of Coloring Maps in Four Colors, *Ann. Math.*, vol. 28, pp. 1–15, pp. 477–492, 1926–1927.

Rill, M.: "Remarks on the Four Color Problem," Preprint no. 94, University of Oklahoma, 1967.

————: "Nine Rings," Ph.D. Thesis, University of Oklahoma, 1974, in preparation.

Ringel, G.: "Farbungsprobleme auf Flächen und Graphen," Berlin, 1959.

————: A Six-Color Problem on the Sphere (German), *Abh. Math. Sem. Univ. Hamburg*, vol. 29, pp. 107–117, 1965.

————: Genus of Graphs, in "Combinatorial Structures and Their Applications" (R. Guy, H. Hanani, N. Sauer, and J. Schonheim, Eds.), Gordon and Breach, New York, 1970, pp. 361–366.

————: "Map Color Theorem," Springer-Verlag, New York, 1974.

——— and J. W. T. Youngs: Solution of the Heawood Map Coloring Problem, *Proc. Nat'l Acad. Sci.*, vol. 60, pp. 438–445, 1968.

——— and ———: Remarks on the Heawood Conjecture, in "Proof Techniques in Graph Theory," (F. Harary, Ed.), Academic Press, New York, 1969a.

——— and ———: Solution of the Heawood Map Coloring Problem, Case 2, *J. Comb. Th.*, vol. 7, pp. 342–353, 1969b.

——— and ———: Solution of the Heawood Map Coloring Problem, Case 8, *J. Comb. Th.*, vol. 7, pp. 353–363, 1969c.

——— and ———: Solution of the Heawood Map Coloring Problem, Case 11, *J. Comb. Th.*, vol. 7, pp. 71–93, 1969d.

Robert, P.: On the Four-Colour Problem, *C. R. Acad. Sci. Paris*, ser. A, vol. 269, pp. 937–939, 1969.

Roberts, F. S., and J. H. Spencer: A Characterization of Clique Graphs, *J. Comb. Th.*, vol. 10, pp. 102–108, 1971.

Rooij, A. van, and H. S. Wilf: The Interchange Graphs of a Finite Graph, *Acta Math. Sci. Hungary*, vol. 16, pp. 263–269, 1965.

Rosa, A: On the Chromatic Number of Steiner Triple Systems, in "Combinatorial Structures and Their Applications," Gordon and Breach, New York, 1970, pp. 369–371.

Rosenfeld, M.: On Tait Colorings of Cubic Graphs, in "Combinatorial Structures and Their Applications," Gordon and Breach, New York, 1970, p. 373.

Rosenstiehl, P. (Ed.): "Theorie des Graphes, Rome I.C.C.," Dunod, Paris, 1967.

Rota, G. C.: On the Foundations of Combinatorial Theory: Theory of Mobius Functions, 2, *Wahrscheinlichkeitstheorie und Verw. Gebiete*, vol. 2, pp. 240–368, 1964.

Roy, B.: Nombre Chromatique et plus Longs Chemins d'un Graphe, *Revue AFIRO*, vol. 1, pp. 127–132, 1967.

———: "Algebre Moderne et Théorie des Graphs," Dunod, two volumes, 1969 and 1970.

Saaty, T. L.: Remarks on the Four Color Problem, the Kempe Catastrophe, *Math. Mag.*, vol. 40, pp. 31–36, 1967.

———: Thirteen Colorful Variations on Guthrie's Four-Color Conjecture, *Am. Math. Monthly*, vol. 79, pp. 2–43, 1972.

Sachs, H.: "Einführung in die Theorie der Endlichen Graphen," Teubner, Leipzig, 1970.

——— and M. Schäuble: Über die Konstruktion von Graphen mit Gewissen Färbungseigenschaften, in "Beiträge zur Graphentheorie," Teubner, Leipzig, 1968, pp. 131–135.

———, H. J. Voss, and H. Walther (Eds.): "Beiträge zur Graphentheorie," Teubner, Leipzig, 1968.

Sauvé, Leopold: On Chromatic Graphs, *Am. Math. Monthly*, vol. 68, pp. 107–111, 1961.

Schäuble, M.: Bemerkungen zu einem Kantenfärbungsproblem, in "Beiträge zur Graphentheorie," Teubner, Leipzig, 1968, pp. 137–142.

Scheim, David E.: The Number of Edge 3-Colorings of a Planar Cubic Graph as a Permanent, *Discrete Math.*, vol. 8, pp. 377–382, 1974.

Sedláček, J.: Some Properties of Interchange Graphs, in "Theory of Graphs and Its Applications," (M. Fiedler, Ed.), Symposium Smolenice, Prague, 1962, reprinted by Academic Press, New York, 1962, pp. 145–150.

Seinsche, D.: On a Property of the Class of *n*-Colorable Graphs, *J. Comb. Th.*, ser. B, vol. 16, pp. 191–193, 1974.

Shannon, C. E.: A Theorem on Coloring the Lines of a Network, *J. Math. Phys.*, vol. 28, pp. 148–151, 1949.

Sheng, T. K.: Graph Extension Preserving Chromatic Number, *Am. Math. Monthly*, vol. 74, p. 844, 1967.

Simmons, G. J.: On a Problem of Erdös Concerning a 3-Coloring of the Unit Sphere, *Discrete Math.*, vol. 8, pp. 81–84, 1974.

Sobczyk, A.: Graph-Coloring and Combinatorial Numbers, *Canad. J. Math.*, vol. 20, pp. 520–534, 1968.

Steen, L. A.: Solution of the Four Color Problem, *Math. Mag.*, vol. 49, pp. 219–222, September, 1976.

Stein, S. K.: Convex Maps, *Proc. Am. Math. Soc.*, vol. 2, pp. 464–466, 1951.

———: *B*-Sets and Planar Maps, *Pacific J. Math.*, vol. 37, pp. 217–224, 1971.

Steinitz, E. (with H. Rademacher): "Vorlesunger über die Theorie der Polyeder," Springer-Verlag, Berlin, 1934.

Stewart, B. M.: On a Theorem of Nordhaus and Gaddum, *J. Comb. Th.*, vol. 6, pp. 217–218, 1969.

Story, W. E.: Note on Mr. Kempe's Paper on the Geographical Problem of the Four Colours, *Am. Jour. Math.*, vol. 2, pp. 201–204, 1879.

Stromquist, W.: Four Color Theorem for Small Maps, *J. Comb. Th.*, 1974a, submitted.

———: Ph.D. Thesis, Harvard University, 1974b, in preparation.

Suranyi, L.: The Covering of Graphs by Cliques, *Studia Sci. Math. Hungary.*, vol. 3, pp. 345–349, 1968.

Synge, J. L.: Two Isomorphs of the Four-Color Conjecture, *Canad. J. Math.*, vol. 19, pp. 1084–1091, 1967.

Szekeres, G.: Polyhedral Decompositions of Cubic Graphs, *Bull. Austral., Math. Soc.,* vol. 8, 1973.

——— and H. S. Wilf: An Inequality for the Chromatic Number of a Graph, *J. Comb. Th.*, vol. 4, pp. 1–3, 1968.

Tait, P. G.: Remarks on the Colouring of Maps, *Proc. Roy. Soc., Edinburgh*, vol. 10, p. 729, 1880a.

———: Note on a Theorem in Geometry of Position, *Trans. Roy. Soc., Edinburgh*, vol. 29, pp. 657–660, 1880b.

———: On Listing's "Topologie," *Phil. Mag.*, vol. 17, pp. 30–46, 1884.

Terry, C. M., L. R. Welch, and J. W. T. Youngs: The Genus of K_{12s}, *J. Comb. Th.*, vol. 2, pp. 43–60, 1967.

———, ———, and ———: Solution of the Heawood Map-Colouring Problem—case 4, *J. Comb. Th.*, vol. 8, pp. 170–174, 1970.

Thomas, J. M.: "The Four Colour Theorem," 60 Slocum Street, Philadelphia, Pennsylvania, 1969 (privately printed).

Toft, Bjarne: "Some Contributions to the Theory of Colour-Critical Graphs," Various Publications Series no. 14, Matematisk Institut Aarhus Universitei, Denmark, 1970.

Tomescu, I.: Sur le Problème du Coloriage des Graphes Généralisés, *C. R. Acad. Sci., Paris*, vol. 267, pp. 250–252, 1968.

Tutte, W. T.: On Hamiltonian Circuits, *J. London Math. Soc.*, vol. 21, pp. 98–101, 1946.

———: The Factorizations of Linear Graphs, *J. London Math. Soc.*, vol. 22, pp. 107–111, 1947.

———: On the Four-Color Conjecture, *Proc. London Math. Soc.*, ser. 2, vol. 50, pp. 137–149, 1948.

———: The Factors of Graphs, *Canad. J. Math.*, vol. 4, p. 314, 1952.

———: A Theorem on Planar Graphs, *Trans Am. Math. Soc.*, vol. 82, pp. 99–116, 1956.

———: A Non-Hamiltonian Graph, *Canad. Math. Bull.*, vol. 3, pp. 1–5, 1960.

———: A Theory of 3-Connected Graphs, *Indag. Math.*, vol. 23, pp. 441–455, 1961.

———: How to Draw a Graph, *Proc. London Math. Soc.*, vol. 13, pp. 743–767, 1963.

———: Lectures on Matroids, *J. Res. Nat. Bur. Stand. Sect.*, ser. B, vol. 69, pp. 1–47, 1965.

———: On the Algebraic Theory of Graph Colorings, *J. Comb. Th.*, vol. 1, pp. 15–50, 1966.

———: "The Connectivity of Graphs," Toronto University Press, Toronto, 1967a.

———: "A Geometrical Version of the Four Color Problem: Proceedings of the Conference held at the University of North Carolina, Chapel Hill, April 10–14, 1967" (R. Bose and T. Dowling, Eds.), University of North Carolina, 1967b.

———: Even and Odd 4-Colorings, in "Proof Techniques in Graph Theory," Academic Press, New York, 1969a.

———: More about Chromatic Polynomials and the Golden Ratio, in "Combinatorial Structures and Their Applications," Gordon and Breach, New York, 1969b, p. 439.

———: Projective Geometry and the 4-Color Problem, in "Recent Progress in Combinatorics," Academic Press, 1969c.

———: On the Enumeration of Four-Colored Maps, *SIAM J. App. Math.*, pp. 454–460, March 1969d.

———: The Golden Ratio in the Theory of Chromatic Polynomials, *Ann. N.Y. Acad. Sci.*, vol. 175, pp. 391–402, 1970a.

———: On Chromatic Polynomials and the Golden Ratio, *J. Comb. Th.*, vol. 9, pp. 289–296, 1970b.

———: "Chromials," Preprint, University of Waterloo, 1971.

———: Non-Hamiltonian Planar Maps, in "Graph Theory and Computing," Academic Press, 1972, pp. 295–301.

————: "Chromatic Sums for Planar Triangulations I: The Case $\lambda = 1$," Research Report CORR 72–2, University of Waterloo, 1972a.

————: "Chromatic Sums for Planar Triangulations IV: The Case $\lambda = \infty$," Research Report CORR 72–4, University of Waterloo, 1972b.

————: "Shimamoto's Attack on the Four Colour Problem," Research Report, University of Waterloo, 1972c.

————: "Codichromatic Graphs," Research Report CORR 73–22, University of Waterloo, 1973a.

————: What is a Map?, in "New Directions in the Theory of Graphs," Academic Press, 1973b.

————: Codichromatic Graphs, J. Comb. Th., vol. 16, pp. 168–174, 1974.

————: Chromials, Chapter in "Studies in Graph Theory," pt. II (D. R. Fulkerson, Ed.), MAA Studies in Mathematics, vol. 12, 1975.

Vallée-Poussin, C. J. De La: Probléme des Quatres Couleurs, Intermédiare des Mathematiciens, vol. 3, pp. 179–180, 1896.

Veblen, O.: An Application of Modular Equations in Analysis Situs, Ann. Math., vol. 14, pp. 86–94, 1912–1913.

————: Analysis Situs, Am. Math. Soc. Colloq. Publ., vol. 5, Cambridge, 1922 (2d ed., New York, 1931).

Vermes, P.: Fixed Points in Graph-Coloring, Magyar Tud. Akad. Mat. Kot. Int. Kozl., vol. 6, pp. 89–96, 1961.

Vigneron, L.: On the Four Colour Problem; Theory of Combination, C. R. Acad. Sci., Paris, vol. 223, pp. 705–707, 1946.

————: Coloration des Réseaux Cubiques, C. R. Acad. Sci., Paris, vol. 223, pp. 705 and 770, 1946; vol. 249, p. 2462, 1959.

————: Sur le Nombre des Composantes de Tait Coupant au Contour Ferme Trace sur un Graphe Cubique Colore Associe au Probleme des Quatre Couleurs, C. R. Acad. Sci., Paris, vol. 253, pp. 43–45, 1961.

Vizing, V. G.: On an Estimate of the Chromatic Class of a p-Graph (Russian), Diskret. Analiz., vol. 3, pp. 25–30, 1964.

————: Critical Graphs with a Given Chromatic Class (Russian), Diskret. Anal., vol. 5, pp. 9–17, 1965a.

————: On Chromatic Class (Russian), Cybernetika, vol. 3, pp. 29–39, 1965b.

————: Chromatic Class of Multigraph, Théorie des Graphs, Journée Internationales d'Etude, Rome, vol. 3, pp. 29–33, 1966.

————: On the Number of Edges in a Graph with Given Radius (Russian), Dokl. Akad. Nauk. SSSR, vol. 173, pp. 1245–1246, 1967.

Wagner, K.: Ein Satz uber Komplexe, Jber Deutsch Math.-Verein, vol. 46, pp. 21–22, 1936a.

————: Bemerkungen zum Vierfarbenproblem, Jber Deutsch Math.-Verein, vol. 46, pp. 26–32, 1936b.

————: Über eine Eigenschaft der ebenen Komplexe, Math. Ann., vol. 114, pp. 570–590, 1937.

————: Beweis einer Abschwächung der Hadwiger-Vermutung, Math. Ann., vol. 153, pp. 139–141, 1964.

Walsh, T., and A. B. Lehman: Counting Rooted Maps by Genus I, II, J. Comb. Th., ser. B, vol. 13, pp. 122–141 and 192–218, 1972.

Walther, H.: "Über die Zerlegung des vollständigen Graphen in paare planare Graphen," Beiträge zum Graphentheorie, H. Sachs, H.-J. Voss, and H. Walther (Eds.), B. G. Teubner Verlag, Leipzig, pp. 189–205, 1968.

Watkins, M. E.: Addendum to Tutte (1967b), pp. 559–560.

————: A Theorem on Tait Colorings with an Application to the Generalized Petersen Graphs, J. Comb. Th., vol. 6, 1969.

Weinberg, L.: Planar Graphs and Matroids, in "Graph Theory and Applications," Springer Lecture Notes, no. 303, Springer-Verlag, 1972, pp. 313–329.

Weinberger, D. B.: Network Flows, Minimum Coverings, and the Four-Colour Conjecture, Op. Res., vol. 24, pp. 272–290, 1976.

Welsh, D. J., and M. B. Powell: An Upper Bound for the Chromatic Number of a Graph and its Application to Time-Tabling Problems, Computer Journal, vol. 10, pp. 85–86, 1967.

Wernicke, P.: Über den Kartographischen Viefarbensatz, Math. Ann., vol. 58, pp. 413–426, 1904.

White, A. T.: "Graphs, Groups and Surfaces," North-Holland, Amsterdam, 1973a.

————: "Orientable Imbeddings of Cayley Graphs," Mathematics Report no. 36, Western Michigan University, 1973b.

Whitney, H.: A Theorem on Graphs, *An. Math.*, vol. 32, pp. 378–390, 1931.

————: Congruent Graphs and the Connectivity of Graphs, *Am. J. Math.*, vol. 54, pp. 150–168, 1932a.

————: A Logical Expansion in Mathematics, *Bull. Am. Math. Soc.*, vol. 38, pp. 572–579, 1932b.

————: Non-Separable and Planar Graphs, *Trans Am. Math. Soc.*, vol. 34, pp. 339–362, 1932c.

————: The Coloring of Graphs, *Ann. Math.*, vol. 33, pp. 688–718, 1932d.

————: A Set of Topological Invariants for Graphs, *Am. J. Math.*, vol. 55, pp. 231–235, 1933a.

————: Isomorphic Graphs, *Am. J. Math.*, vol. 55, pp. 245–254, 1933b.

————: Planar Graphs, *Fund. Math.*, vol. 21, pp. 73–84, 1933c.

————: On the Abstract Properties of Linear Dependence, *Am. J. Math.*, vol. 57, pp. 509–533, 1935.

————, and W. T. Tutte: Kempe Chains and the Four Color Problem, *Utilitas Math.*, vol. 2, pp. 241–281, 1972. Also Chapter in Studies in "Graph Theory," pt. II (D. R. Fulkerson, Ed.), MAA Studies in Mathematics, vol. 13, 1975.

————: A Numerical Equivalent of the Four Color Map Problem, *Monatshefte fur Mathematik und Physik*, pp. 207–213, 1937.

Wilf, H. S.: The Eigenvalues of a Graph and its Chromatic Number, *J. London Math. Soc.*, vol. 42, pp. 330–332, 1967.

————: Hadamard Determinants, Mobius Functions and the Chromatic Number of a Graph, *Bull. Am. Math. Soc.*, pp. 960–964, 1968.

Williams, M. R.: "A Graph Theory Model for the Computer Solution of University Time-Tables and Related Problems," Ph.D. Thesis, University of Glasgow, 1969.

————: The Coloring of Very Large Graphs, in "Combinatorial Structures and their Applications," Gordon and Breach, 1970.

Wilson, R. J.: "Introduction to Graph Theory," Academic Press, New York, 1972.

Winn, C. E.: A Case of Coloration in the Four Color Problem, *Am. J. of Math.*, vol. 59, pp. 515–528, 1937.

————: On Certain Reductions in the Four Color Problem, *J. Math. Phys.*, vol. 16, pp. 159–171, 1938.

————: Sur l'Historique du Problème des Quatres Couleurs, *Bull. Inst. Egypte*, vol. 20, pp. 191–192, 1939.

————: On the Minimum Number of Polygons in an Irreducible Map, *Am. J. Math.*, vol. 62, pp. 406–416, 1940.

Woodall, D. R.: Property *B* and the Four-Colour Problem, in "Combinatorics, Proceedings of the 1972 Oxford Combinatorial Conference" (D. J. A. Welsh and D. R. Woodall, Eds.), 1972, pp. 322–340.

Wright, E.: Counting Colored Graphs, *Canad. J. Math.*, vol. 13, pp. 683–693, 1961.

Yamabe, H., and D. Pope: A Computational Approach to the Four-Color Problem, *Math. Comp.*, pp. 250–253, 1961.

Young, H. P.: A Quick Proof of Wagner's Equivalence Theorem, *J. London Math. Soc.*, ser. 2, vol. 3, pp. 661–664, 1971.

Youngs, J. W. T.: Minimal Imbeddings and the Genus of a Graph, *J. Math. Mech.*, vol. 12, pp. 303–315, 1963.

————: The Heawood Map Coloring Conjecture, Chapter 12 in "Graph Theory and Theoretical Physics" (F. Harary, Ed.), Academic Press, London, 1967.

————: The Heawood Map-Coloring Problem—Cases 1, 7 and 10, *J. Comb. Th.*, vol. 8, pp. 220–231, 1970a.

————: The Mystery of the Heawood Conjecture, in "Graph Theory and Its Applications" (B. Harris, Ed.), Academic Press, New York, 1970b, pp. 17–50.

————Solution of the Heawood Map-Coloring Problem—Cases 3, 5, 6 and 9, *J. Comb. Th.*, vol. 8, pp. 175–219, 1970c.

————: Remarks on the Four Color Problem, in "Combinatorial Structures and Their Applications," Gordon and Breach, New York, 1970d.

Zeeman, E. C.: "Seminar on Piecewise-Linear Topology" (mimeographed), Inst. des Hautes Études Scientifique, 1965.

Zeidl, B.: Über 4-und 5-Chrome Graphen, *J. London Math. Soc.*, vol. 27, pp. 85–92, 1952.

Zykov, A. A.: "On Some Properties of Linear Complexes" (Russian), *Mat. Sbolnik*, pp. 163–168, 1949. American Mathematical Society Translation no. 79, 1952.

———: "Theory of Finite Graphs" (Russian), Nauka, Novosibirsk, 1969a.

———: On a Vector Space Associated with Hadwiger's Hypothesis, *Kokl. Akad. Nauk. SSSR*, vol. 187, no. 6, 1969; *Soviet Math. Dokl.*, vol. 10, no. 4, p. 1023, 1969b.

SUPPLEMENTARY BIBLIOGRAPHY

Appel, K., and W. Haken: The Solution of the Four-Color Map Problem, *Scientific American,* October 1977, pp. 108–121.

Bernhart, F.: A Digest of the Four Color Theorem, *J. of Graph Theory,* vol. 1, pp. 207–225, 1977.

Haken, W.: An Attempt to Understand the Four Color Problem, *J. of Graph Theory,* vol. 1, pp. 193–206, 1977.

Kainen, P. C.: The Significance of the Four Color Problem, in "Proc. West Coast Conference on Combinatorics, Graph Theory, and Computing," Utilitas Math. Publ., Winnipeg, 1979, pp. 49–66.

Saaty, T. L.: Graph Theory: The Four Color Problem, in "McGraw-Hill Yearbook of Science & Technology," McGraw-Hill, New York, 1981.

INDEX

A CATALOG OF SELECTED
DOVER BOOKS
IN SCIENCE AND MATHEMATICS

A CATALOG OF SELECTED
DOVER BOOKS
IN SCIENCE AND MATHEMATICS

QUALITATIVE THEORY OF DIFFERENTIAL EQUATIONS, V.V. Nemytskii and V.V. Stepanov. Classic graduate-level text by two prominent Soviet mathematicians covers classical differential equations as well as topological dynamics and ergodic theory. Bibliographies. 523pp. 5⅜ × 8½.   65954-2 Pa. $10.95

MATRICES AND LINEAR ALGEBRA, Hans Schneider and George Phillip Barker. Basic textbook covers theory of matrices and its applications to systems of linear equations and related topics such as determinants, eigenvalues and differential equations. Numerous exercises. 432pp. 5⅜ × 8½.   66014-1 Pa. $8.95

QUANTUM THEORY, David Bohm. This advanced undergraduate-level text presents the quantum theory in terms of qualitative and imaginative concepts, followed by specific applications worked out in mathematical detail. Preface. Index. 655pp. 5⅜ × 8½.   65969-0 Pa. $12.95

ATOMIC PHYSICS (8th edition), Max Born. Nobel laureate's lucid treatment of kinetic theory of gases, elementary particles, nuclear atom, wave-corpuscles, atomic structure and spectral lines, much more. Over 40 appendices, bibliography. 495pp. 5⅜ × 8½.   65984-4 Pa. $11.95

ELECTRONIC STRUCTURE AND THE PROPERTIES OF SOLIDS: The Physics of the Chemical Bond, Walter A. Harrison. Innovative text offers basic understanding of the electronic structure of covalent and ionic solids, simple metals, transition metals and their compounds. Problems. 1980 edition. 582pp. 6⅛ × 9¼.   66021-4 Pa. $14.95

BOUNDARY VALUE PROBLEMS OF HEAT CONDUCTION, M. Necati Özisik. Systematic, comprehensive treatment of modern mathematical methods of solving problems in heat conduction and diffusion. Numerous examples and problems. Selected references. Appendices. 505pp. 5⅜ × 8½.   65990-9 Pa. $11.95

A SHORT HISTORY OF CHEMISTRY (3rd edition), J.R. Partington. Classic exposition explores origins of chemistry, alchemy, early medical chemistry, nature of atmosphere, theory of valency, laws and structure of atomic theory, much more. 428pp. 5⅜ × 8½. (Available in U.S. only)   65977-1 Pa. $10.95

A HISTORY OF ASTRONOMY, A. Pannekoek. Well-balanced, carefully reasoned study covers such topics as Ptolemaic theory, work of Copernicus, Kepler, Newton, Eddington's work on stars, much more. Illustrated. References. 521pp. 5⅜ × 8½.   65994-1 Pa. $11.95

PRINCIPLES OF METEOROLOGICAL ANALYSIS, Walter J. Saucier. Highly respected, abundantly illustrated classic reviews atmospheric variables, hydrostatics, static stability, various analyses (scalar, cross-section, isobaric, isentropic, more). For intermediate meteorology students. 454pp. 6⅛ × 9¼. 65979-8 Pa. $12.95

RELATIVITY, THERMODYNAMICS AND COSMOLOGY, Richard C. Tolman. Landmark study extends thermodynamics to special, general relativity; also applications of relativistic mechanics, thermodynamics to cosmological models. 501pp. 5⅜ × 8½. 65383-8 Pa. $12.95

APPLIED ANALYSIS, Cornelius Lanczos. Classic work on analysis and design of finite processes for approximating solution of analytical problems. Algebraic equations, matrices, harmonic analysis, quadrature methods, much more. 559pp. 5⅜ × 8½. 65656-X Pa. $12.95

SPECIAL RELATIVITY FOR PHYSICISTS, G. Stephenson and C.W. Kilmister. Concise elegant account for nonspecialists. Lorentz transformation, optical and dynamical applications, more. Bibliography. 108pp. 5⅜ × 8½. 65519-9 Pa. $4.95

INTRODUCTION TO ANALYSIS, Maxwell Rosenlicht. Unusually clear, accessible coverage of set theory, real number system, metric spaces, continuous functions, Riemann integration, multiple integrals, more. Wide range of problems. Undergraduate level. Bibliography. 254pp. 5⅜ × 8½. 65038-3 Pa. $7.95

INTRODUCTION TO QUANTUM MECHANICS With Applications to Chemistry, Linus Pauling & E. Bright Wilson, Jr. Classic undergraduate text by Nobel Prize winner applies quantum mechanics to chemical and physical problems. Numerous tables and figures enhance the text. Chapter bibliographies. Appendices. Index. 468pp. 5⅜ × 8½. 64871-0 Pa. $10.95

ASYMPTOTIC EXPANSIONS OF INTEGRALS, Norman Bleistein & Richard A. Handelsman. Best introduction to important field with applications in a variety of scientific disciplines. New preface. Problems. Diagrams. Tables. Bibliography. Index. 448pp. 5⅜ × 8½. 65082-0 Pa. $11.95

MATHEMATICS APPLIED TO CONTINUUM MECHANICS, Lee A. Segel. Analyzes models of fluid flow and solid deformation. For upper-level math, science and engineering students. 608pp. 5⅜ × 8½. 65369-2 Pa. $13.95

ELEMENTS OF REAL ANALYSIS, David A. Sprecher. Classic text covers fundamental concepts, real number system, point sets, functions of a real variable, Fourier series, much more. Over 500 exercises. 352pp. 5⅜ × 8½. 65385-4 Pa. $9.95

PHYSICAL PRINCIPLES OF THE QUANTUM THEORY, Werner Heisenberg. Nobel Laureate discusses quantum theory, uncertainty, wave mechanics, work of Dirac, Schroedinger, Compton, Wilson, Einstein, etc. 184pp. 5⅜ × 8½. 60113-7 Pa. $4.95

INTRODUCTORY REAL ANALYSIS, A.N. Kolmogorov, S.V. Fomin. Translated by Richard A. Silverman. Self-contained, evenly paced introduction to real and functional analysis. Some 350 problems. 403pp. 5⅜ × 8½. 61226-0 Pa. $8.95

PROBLEMS AND SOLUTIONS IN QUANTUM CHEMISTRY AND PHYSICS, Charles S. Johnson, Jr. and Lee G. Pedersen. Unusually varied problems, detailed solutions in coverage of quantum mechanics, wave mechanics, angular momentum, molecular spectroscopy, scattering theory, more. 280 problems plus 139 supplementary exercises. 430pp. 6½ × 9¼. 65236-X Pa. $11.95

ASYMPTOTIC METHODS IN ANALYSIS, N.G. de Bruijn. An inexpensive, comprehensive guide to asymptotic methods—the pioneering work that teaches by explaining worked examples in detail. Index. 224pp. 5⅜ × 8½.   64221-6 Pa. $5.95

OPTICAL RESONANCE AND TWO-LEVEL ATOMS, L. Allen and J.H. Eberly. Clear, comprehensive introduction to basic principles behind all quantum optical resonance phenomena. 53 illustrations. Preface. Index. 256pp. 5⅜ × 8½.
65533-4 Pa. $7.95

COMPLEX VARIABLES, Francis J. Flanigan. Unusual approach, delaying complex algebra till harmonic functions have been analyzed from real variable viewpoint. Includes problems with answers. 364pp. 5⅜ × 8½.   61388-7 Pa. $7.95

ATOMIC SPECTRA AND ATOMIC STRUCTURE, Gerhard Herzberg. One of best introductions; especially for specialist in other fields. Treatment is physical rather than mathematical. 80 illustrations. 257pp. 5⅜ × 8½.   60115-3 Pa. $4.95

APPLIED COMPLEX VARIABLES, John W. Dettman. Step-by-step coverage of fundamentals of analytic function theory—plus lucid exposition of five important applications: Potential Theory; Ordinary Differential Equations; Fourier Transforms; Laplace Transforms; Asymptotic Expansions. 66 figures. Exercises at chapter ends. 512pp. 5⅜ × 8½.   64670-X Pa. $10.95

ULTRASONIC ABSORPTION: An Introduction to the Theory of Sound Absorption and Dispersion in Gases, Liquids and Solids, A.B. Bhatia. Standard reference in the field provides a clear, systematically organized introductory review of fundamental concepts for advanced graduate students, research workers. Numerous diagrams. Bibliography. 440pp. 5⅜ × 8½.   64917-2 Pa. $11.95

UNBOUNDED LINEAR OPERATORS: Theory and Applications, Seymour Goldberg. Classic presents systematic treatment of the theory of unbounded linear operators in normed linear spaces with applications to differential equations. Bibliography. 199pp. 5⅜ × 8½.   64830-3 Pa. $7.00

LIGHT SCATTERING BY SMALL PARTICLES, H.C. van de Hulst. Comprehensive treatment including full range of useful approximation methods for researchers in chemistry, meteorology and astronomy. 44 illustrations. 470pp. 5⅜ × 8½.   64228-3 Pa. $9.95

CONFORMAL MAPPING ON RIEMANN SURFACES, Harvey Cohn. Lucid, insightful book presents ideal coverage of subject. 334 exercises make book perfect for self-study. 55 figures. 352pp. 5⅜ × 8¼.   64025-6 Pa. $8.95

OPTICKS, Sir Isaac Newton. Newton's own experiments with spectroscopy, colors, lenses, reflection, refraction, etc., in language the layman can follow. Foreword by Albert Einstein. 532pp. 5⅜ × 8½.   60205-2 Pa. $8.95

GENERALIZED INTEGRAL TRANSFORMATIONS, A.H. Zemanian. Graduate-level study of recent generalizations of the Laplace, Mellin, Hankel, K. Weierstrass, convolution and other simple transformations. Bibliography. 320pp. 5⅜ × 8½.   65375-7 Pa. $7.95

THE ELECTROMAGNETIC FIELD, Albert Shadowitz. Comprehensive undergraduate text covers basics of electric and magnetic fields, builds up to electromagnetic theory. Also related topics, including relativity. Over 900 problems. 768pp. 5⅜ × 8¼. 65660-8 Pa. $15.95

FOURIER SERIES, Georgi P. Tolstov. Translated by Richard A. Silverman. A valuable addition to the literature on the subject, moving clearly from subject to subject and theorem to theorem. 107 problems, answers. 336pp. 5⅜ × 8½. 63317-9 Pa. $7.95

THEORY OF ELECTROMAGNETIC WAVE PROPAGATION, Charles Herach Papas. Graduate-level study discusses the Maxwell field equations, radiation from wire antennas, the Doppler effect and more. xiii + 244pp. 5⅜ × 8½. 65678-0 Pa. $6.95

DISTRIBUTION THEORY AND TRANSFORM ANALYSIS: An Introduction to Generalized Functions, with Applications, A.H. Zemanian. Provides basics of distribution theory, describes generalized Fourier and Laplace transformations. Numerous problems. 384pp. 5⅜ × 8½. 65479-6 Pa. $9.95

THE PHYSICS OF WAVES, William C. Elmore and Mark A. Heald. Unique overview of classical wave theory. Acoustics, optics, electromagnetic radiation, more. Ideal as classroom text or for self-study. Problems. 477pp. 5⅜ × 8½. 64926-1 Pa. $10.95

CALCULUS OF VARIATIONS WITH APPLICATIONS, George M. Ewing. Applications-oriented introduction to variational theory develops insight and promotes understanding of specialized books, research papers. Suitable for advanced undergraduate/graduate students as primary, supplementary text. 352pp. 5⅜ × 8½. 64856-7 Pa. $8.50

A TREATISE ON ELECTRICITY AND MAGNETISM, James Clerk Maxwell. Important foundation work of modern physics. Brings to final form Maxwell's theory of electromagnetism and rigorously derives his general equations of field theory. 1,084pp. 5⅜ × 8½. 60636-8, 60637-6 Pa., Two-vol. set $19.90

AN INTRODUCTION TO THE CALCULUS OF VARIATIONS, Charles Fox. Graduate-level text covers variations of an integral, isoperimetrical problems, least action, special relativity, approximations, more. References. 279pp. 5⅜ × 8½. 65499-0 Pa. $7.95

HYDRODYNAMIC AND HYDROMAGNETIC STABILITY, S. Chandrasekhar. Lucid examination of the Rayleigh-Benard problem; clear coverage of the theory of instabilities causing convection. 704pp. 5⅜ × 8¼. 64071-X Pa. $12.95

CALCULUS OF VARIATIONS, Robert Weinstock. Basic introduction covering isoperimetric problems, theory of elasticity, quantum mechanics, electrostatics, etc. Exercises throughout. 326pp. 5⅜ × 8½. 63069-2 Pa. $7.95

DYNAMICS OF FLUIDS IN POROUS MEDIA, Jacob Bear. For advanced students of ground water hydrology, soil mechanics and physics, drainage and irrigation engineering and more. 335 illustrations. Exercises, with answers. 784pp. 6⅛ × 9¼. 65675-6 Pa. $19.95

NUMERICAL METHODS FOR SCIENTISTS AND ENGINEERS, Richard Hamming. Classic text stresses frequency approach in coverage of algorithms, polynomial approximation, Fourier approximation, exponential approximation, other topics. Revised and enlarged 2nd edition. 721pp. 5⅜ × 8½.
65241-6 Pa. $14.95

THEORETICAL SOLID STATE PHYSICS, Vol. I: Perfect Lattices in Equilibrium; Vol. II: Non-Equilibrium and Disorder, William Jones and Norman H. March. Monumental reference work covers fundamental theory of equilibrium properties of perfect crystalline solids, non-equilibrium properties, defects and disordered systems. Appendices. Problems. Preface. Diagrams. Index. Bibliography. Total of 1,301pp. 5⅜ × 8½. Two volumes. Vol. I 65015-4 Pa. $12.95
Vol. II 65016-2 Pa. $12.95

OPTIMIZATION THEORY WITH APPLICATIONS, Donald A. Pierre. Broad-spectrum approach to important topic. Classical theory of minima and maxima, calculus of variations, simplex technique and linear programming, more. Many problems, examples. 640pp. 5⅜ × 8½. 65205-X Pa. $13.95

THE MODERN THEORY OF SOLIDS, Frederick Seitz. First inexpensive edition of classic work on theory of ionic crystals, free-electron theory of metals and semiconductors, molecular binding, much more. 736pp. 5⅜ × 8½.
65482-6 Pa. $15.95

ESSAYS ON THE THEORY OF NUMBERS, Richard Dedekind. Two classic essays by great German mathematician: on the theory of irrational numbers; and on transfinite numbers and properties of natural numbers. 115pp. 5⅜ × 8½.
21010-3 Pa. $4.95

THE FUNCTIONS OF MATHEMATICAL PHYSICS, Harry Hochstadt. Comprehensive treatment of orthogonal polynomials, hypergeometric functions, Hill's equation, much more. Bibliography. Index. 322pp. 5⅜ × 8½. 65214-9 Pa. $9.95

NUMBER THEORY AND ITS HISTORY, Oystein Ore. Unusually clear, accessible introduction covers counting, properties of numbers, prime numbers, much more. Bibliography. 380pp. 5⅜ × 8½. 65620-9 Pa. $8.95

THE VARIATIONAL PRINCIPLES OF MECHANICS, Cornelius Lanczos. Graduate level coverage of calculus of variations, equations of motion, relativistic mechanics, more. First inexpensive paperbound edition of classic treatise. Index. Bibliography. 418pp. 5⅜ × 8½. 65067-7 Pa. $10.95

MATHEMATICAL TABLES AND FORMULAS, Robert D. Carmichael and Edwin R. Smith. Logarithms, sines, tangents, trig functions, powers, roots, reciprocals, exponential and hyperbolic functions, formulas and theorems. 269pp. 5⅜ × 8½. 60111-0 Pa. $5.95

THEORETICAL PHYSICS, Georg Joos, with Ira M. Freeman. Classic overview covers essential math, mechanics, electromagnetic theory, thermodynamics, quantum mechanics, nuclear physics, other topics. First paperback edition. xxiii + 885pp. 5⅜ × 8½. 65227-0 Pa. $18.95

HANDBOOK OF MATHEMATICAL FUNCTIONS WITH FORMULAS, GRAPHS, AND MATHEMATICAL TABLES, edited by Milton Abramowitz and Irene A. Stegun. Vast compendium: 29 sets of tables, some to as high as 20 places. 1,046pp. 8 × 10½. 61272-4 Pa. $21.95

MATHEMATICAL METHODS IN PHYSICS AND ENGINEERING, John W. Dettman. Algebraically based approach to vectors, mapping, diffraction, other topics in applied math. Also generalized functions, analytic function theory, more. Exercises. 448pp. 5⅜ × 8¼. 65649-7 Pa. $8.95

A SURVEY OF NUMERICAL MATHEMATICS, David M. Young and Robert Todd Gregory. Broad self-contained coverage of computer-oriented numerical algorithms for solving various types of mathematical problems in linear algebra, ordinary and partial, differential equations, much more. Exercises. Total of 1,248pp. 5⅜ × 8½. Two volumes. Vol. I 65691-8 Pa. $13.95
Vol. II 65692-6 Pa. $13.95

TENSOR ANALYSIS FOR PHYSICISTS, J.A. Schouten. Concise exposition of the mathematical basis of tensor analysis, integrated with well-chosen physical examples of the theory. Exercises. Index. Bibliography. 289pp. 5⅜ × 8½. 65582-2 Pa. $7.95

INTRODUCTION TO NUMERICAL ANALYSIS (2nd Edition), F.B. Hildebrand. Classic, fundamental treatment covers computation, approximation, interpolation, numerical differentiation and integration, other topics. 150 new problems. 669pp. 5⅜ × 8½. 65363-3 Pa. $14.95

INVESTIGATIONS ON THE THEORY OF THE BROWNIAN MOVEMENT, Albert Einstein. Five papers (1905–8) investigating dynamics of Brownian motion and evolving elementary theory. Notes by R. Fürth. 122pp. 5⅜ × 8½. 60304-0 Pa. $3.95

NUMERICAL METHODS FOR SCIENTISTS AND ENGINEERS, Richard Hamming. Classic text stresses frequency approach in coverage of algorithms, polynomial approximation, Fourier approximation, exponential approximation, other topics. Revised and enlarged 2nd edition. 721pp. 5⅜ × 8½. 65241-6 Pa. $14.95

AN INTRODUCTION TO STATISTICAL THERMODYNAMICS, Terrell L. Hill. Excellent basic text offers wide-ranging coverage of quantum statistical mechanics, systems of interacting molecules, quantum statistics, more. 523pp. 5⅜ × 8½. 65242-4 Pa. $11.95

ELEMENTARY DIFFERENTIAL EQUATIONS, William Ted Martin and Eric Reissner. Exceptionally clear, comprehensive introduction at undergraduate level. Nature and origin of differential equations, differential equations of first, second and higher orders. Picard's Theorem, much more. Problems with solutions. 331pp. 5⅜ × 8½. 65024-3 Pa. $8.95

STATISTICAL PHYSICS, Gregory H. Wannier. Classic text combines thermodynamics, statistical mechanics and kinetic theory in one unified presentation of thermal physics. Problems with solutions. Bibliography. 532pp. 5⅜ × 8½. 65401-X Pa. $10.95

ORDINARY DIFFERENTIAL EQUATIONS, Morris Tenenbaum and Harry Pollard. Exhaustive survey of ordinary differential equations for undergraduates in mathematics, engineering, science. Thorough analysis of theorems. Diagrams. Bibliography. Index. 818pp. 5⅜ × 8½. 64940-7 Pa. $15.95

STATISTICAL MECHANICS: Principles and Applications, Terrell L. Hill. Standard text covers fundamentals of statistical mechanics, applications to fluctuation theory, imperfect gases, distribution functions, more. 448pp. 5⅜ × 8½. 65390-0 Pa. $9.95

ORDINARY DIFFERENTIAL EQUATIONS AND STABILITY THEORY: An Introduction, David A. Sánchez. Brief, modern treatment. Linear equation, stability theory for autonomous and nonautonomous systems, etc. 164pp. 5⅜ × 8¼. 63828-6 Pa. $4.95

THIRTY YEARS THAT SHOOK PHYSICS: The Story of Quantum Theory, George Gamow. Lucid, accessible introduction to influential theory of energy and matter. Careful explanations of Dirac's anti-particles, Bohr's model of the atom, much more. 12 plates. Numerous drawings. 240pp. 5⅜ × 8½. 24895-X Pa. $5.95

ORDINARY DIFFERENTIAL EQUATIONS, I.G. Petrovski. Covers basic concepts, some differential equations and such aspects of the general theory as Euler lines, Arzel's theorem, Peano's existence theorem, Osgood's uniqueness theorem, more. 45 figures. Problems. Bibliography. Index. xi + 232pp. 5⅜ × 8½. 64683-1 Pa. $6.95

GREAT EXPERIMENTS IN PHYSICS: Firsthand Accounts from Galileo to Einstein, edited by Morris H. Shamos. 25 crucial discoveries: Newton's laws of motion, Chadwick's study of the neutron, Hertz on electromagnetic waves, more. Original accounts clearly annotated. 370pp. 5⅜ × 8½. 25346-5 Pa. $8.95

INTRODUCTION TO PARTIAL DIFFERENTIAL EQUATIONS WITH APPLICATIONS, E.C. Zachmanoglou and Dale W. Thoe. Essentials of partial differential equations applied to common problems in engineering and the physical sciences. Problems and answers. 416pp. 5⅜ × 8½. 65251-3 Pa. $9.95

BURNHAM'S CELESTIAL HANDBOOK, Robert Burnham, Jr. Thorough guide to the stars beyond our solar system. Exhaustive treatment. Alphabetical by constellation: Andromeda to Cetus in Vol. 1; Chamaeleon to Orion in Vol. 2; and Pavo to Vulpecula in Vol. 3. Hundreds of illustrations. Index in Vol. 3. 2,000pp. 6⅛ × 9¼. 23567-X, 23568-8, 23673-0 Pa., Three-vol. set $41.85

ASYMPTOTIC EXPANSIONS FOR ORDINARY DIFFERENTIAL EQUATIONS, Wolfgang Wasow. Outstanding text covers asymptotic power series, Jordan's canonical form, turning point problems, singular perturbations, much more. Problems. 384pp. 5⅜ × 8½. 65456-7 Pa. $9.95

AMATEUR ASTRONOMER'S HANDBOOK, J.B. Sidgwick. Timeless, comprehensive coverage of telescopes, mirrors, lenses, mountings, telescope drives, micrometers, spectroscopes, more. 189 illustrations. 576pp. 5⅜ × 8¼. 24034-7 Pa. $9.95

CATALOG OF DOVER BOOKS

SPECIAL FUNCTIONS, N.N. Lebedev. Translated by Richard Silverman. Famous Russian work treating more important special functions, with applications to specific problems of physics and engineering. 38 figures. 308pp. 5⅜ × 8½.
60624-4 Pa. $7.95

OBSERVATIONAL ASTRONOMY FOR AMATEURS, J.B. Sidgwick. Mine of useful data for observation of sun, moon, planets, asteroids, aurorae, meteors, comets, variables, binaries, etc. 39 illustrations. 384pp. 5⅜ × 8¼. (Available in U.S. only)
24033-9 Pa. $5.95

INTEGRAL EQUATIONS, F.G. Tricomi. Authoritative, well-written treatment of extremely useful mathematical tool with wide applications. Volterra Equations, Fredholm Equations, much more. Advanced undergraduate to graduate level. Exercises. Bibliography. 238pp. 5⅜ × 8½.
64828-1 Pa. $6.95

CELESTIAL OBJECTS FOR COMMON TELESCOPES, T.W. Webb. Inestimable aid for locating and identifying nearly 4,000 celestial objects. 77 illustrations. 645pp. 5⅜ × 8½.
20917-2, 20918-0 Pa., Two-vol. set $12.00

MODERN NONLINEAR EQUATIONS, Thomas L. Saaty. Emphasizes practical solution of problems; covers seven types of equations. ". . . a welcome contribution to the existing literature. . . ."—Math Reviews. 490pp. 5⅜ × 8½. 64232-1 Pa. $9.95

FUNDAMENTALS OF ASTRODYNAMICS, Roger Bate et al. Modern approach developed by U.S. Air Force Academy. Designed as a first course. Problems, exercises. Numerous illustrations. 455pp. 5⅜ × 8½.
60061-0 Pa. $8.95

INTRODUCTION TO LINEAR ALGEBRA AND DIFFERENTIAL EQUATIONS, John W. Dettman. Excellent text covers complex numbers, determinants, orthonormal bases, Laplace transforms, much more. Exercises with solutions. Undergraduate level. 416pp. 5⅜ × 8½.
65191-6 Pa. $9.95

INCOMPRESSIBLE AERODYNAMICS, edited by Bryan Thwaites. Covers theoretical and experimental treatment of the uniform flow of air and viscous fluids past two-dimensional aerofoils and three-dimensional wings; many other topics. 654pp. 5⅜ × 8½.
65465-6 Pa. $15.95

INTRODUCTION TO DIFFERENCE EQUATIONS, Samuel Goldberg. Exceptionally clear exposition of important discipline with applications to sociology, psychology, economics. Many illustrative examples; over 250 problems. 260pp. 5⅜ × 8½.
65084-7 Pa. $6.95

LAMINAR BOUNDARY LAYERS, edited by L. Rosenhead. Engineering classic covers steady boundary layers in two- and three-dimensional flow, unsteady boundary layers, stability, observational techniques, much more. 708pp. 5⅜ × 8½.
65646-2 Pa. $15.95

LECTURES ON CLASSICAL DIFFERENTIAL GEOMETRY, Second Edition, Dirk J. Struik. Excellent brief introduction covers curves, theory of surfaces, fundamental equations, geometry on a surface, conformal mapping, other topics. Problems. 240pp. 5⅜ × 8½.
65609-8 Pa. $6.95

ROTARY-WING AERODYNAMICS, W.Z. Stepniewski. Clear, concise text covers aerodynamic phenomena of the rotor and offers guidelines for helicopter performance evaluation. Originally prepared for NASA. 537 figures. 640pp. 6⅛ × 9¼.
64647-5 Pa. $14.95

DIFFERENTIAL GEOMETRY, Heinrich W. Guggenheimer. Local differential geometry as an application of advanced calculus and linear algebra. Curvature, transformation groups, surfaces, more. Exercises. 62 figures. 378pp. 5⅜ × 8½.
63433-7 Pa. $7.95

INTRODUCTION TO SPACE DYNAMICS, William Tyrrell Thomson. Comprehensive, classic introduction to space-flight engineering for advanced undergraduate and graduate students. Includes vector algebra, kinematics, transformation of coordinates. Bibliography. Index. 352pp. 5⅜ × 8½.   65113-4 Pa. $8.95

A SURVEY OF MINIMAL SURFACES, Robert Osserman. Up-to-date, in-depth discussion of the field for advanced students. Corrected and enlarged edition covers new developments. Includes numerous problems. 192pp. 5⅜ × 8½.
64998-9 Pa. $8.95

ANALYTICAL MECHANICS OF GEARS, Earle Buckingham. Indispensable reference for modern gear manufacture covers conjugate gear-tooth action, gear-tooth profiles of various gears, many other topics. 263 figures. 102 tables. 546pp. 5⅜ × 8½.   65712-4 Pa. $11.95

SET THEORY AND LOGIC, Robert R. Stoll. Lucid introduction to unified theory of mathematical concepts. Set theory and logic seen as tools for conceptual understanding of real number system. 496pp. 5⅜ × 8¼.   63829-4 Pa. $8.95

A HISTORY OF MECHANICS, René Dugas. Monumental study of mechanical principles from antiquity to quantum mechanics. Contributions of ancient Greeks, Galileo, Leonardo, Kepler, Lagrange, many others. 671pp. 5⅜ × 8½.
65632-2 Pa. $14.95

FAMOUS PROBLEMS OF GEOMETRY AND HOW TO SOLVE THEM, Benjamin Bold. Squaring the circle, trisecting the angle, duplicating the cube: learn their history, why they are impossible to solve, then solve them yourself. 128pp. 5⅜ × 8½.   24297-8 Pa. $3.95

MECHANICAL VIBRATIONS, J.P. Den Hartog. Classic textbook offers lucid explanations and illustrative models, applying theories of vibrations to a variety of practical industrial engineering problems. Numerous figures. 233 problems, solutions. Appendix. Index. Preface. 436pp. 5⅜ × 8½.   64785-4 Pa. $8.95

CURVATURE AND HOMOLOGY, Samuel I. Goldberg. Thorough treatment of specialized branch of differential geometry. Covers Riemannian manifolds, topology of differentiable manifolds, compact Lie groups, other topics. Exercises. 315pp. 5⅜ × 8½.   64314-X Pa. $8.95

HISTORY OF STRENGTH OF MATERIALS, Stephen P. Timoshenko. Excellent historical survey of the strength of materials with many references to the theories of elasticity and structure. 245 figures. 452pp. 5⅜ × 8½. 61187-6 Pa. $10.95

CATALOG OF DOVER BOOKS

GEOMETRY OF COMPLEX NUMBERS, Hans Schwerdtfeger. Illuminating, widely praised book on analytic geometry of circles, the Moebius transformation, and two-dimensional non-Euclidean geometries. 200pp. 5⅜ × 8¼.

63830-8 Pa. $6.95

MECHANICS, J.P. Den Hartog. A classic introductory text or refresher. Hundreds of applications and design problems illuminate fundamentals of trusses, loaded beams and cables, etc. 334 answered problems. 462pp. 5⅜ × 8½. 60754-2 Pa. $8.95

TOPOLOGY, John G. Hocking and Gail S. Young. Superb one-year course in classical topology. Topological spaces and functions, point-set topology, much more. Examples and problems. Bibliography. Index. 384pp. 5⅜ × 8¼.

65676-4 Pa. $7.95

STRENGTH OF MATERIALS, J.P. Den Hartog. Full, clear treatment of basic material (tension, torsion, bending, etc.) plus advanced material on engineering methods, applications. 350 answered problems. 323pp. 5⅜ × 8½. 60755-0 Pa. $7.50

ELEMENTARY CONCEPTS OF TOPOLOGY, Paul Alexandroff. Elegant, intuitive approach to topology from set-theoretic topology to Betti groups; how concepts of topology are useful in math and physics. 25 figures. 57pp. 5⅜ × 8½.

60747-X Pa. $2.95

ADVANCED STRENGTH OF MATERIALS, J.P. Den Hartog. Superbly written advanced text covers torsion, rotating disks, membrane stresses in shells, much more. Many problems and answers. 388pp. 5⅜ × 8½. 65407-9 Pa. $9.95

COMPUTABILITY AND UNSOLVABILITY, Martin Davis. Classic graduate-level introduction to theory of computability, usually referred to as theory of recurrent functions. New preface and appendix. 288pp. 5⅜ × 8½. 61471-9 Pa. $6.95

GENERAL CHEMISTRY, Linus Pauling. Revised 3rd edition of classic first-year text by Nobel laureate. Atomic and molecular structure, quantum mechanics, statistical mechanics, thermodynamics correlated with descriptive chemistry. Problems. 992pp. 5⅜ × 8½. 65622-5 Pa. $18.95

AN INTRODUCTION TO MATRICES, SETS AND GROUPS FOR SCIENCE STUDENTS, G. Stephenson. Concise, readable text introduces sets, groups, and most importantly, matrices to undergraduate students of physics, chemistry, and engineering. Problems. 164pp. 5⅜ × 8½. 65077-4 Pa. $5.95

THE HISTORICAL BACKGROUND OF CHEMISTRY, Henry M. Leicester. Evolution of ideas, not individual biography. Concentrates on formulation of a coherent set of chemical laws. 260pp. 5⅜ × 8½. 61053-5 Pa. $6.00

THE PHILOSOPHY OF MATHEMATICS: An Introductory Essay, Stephan Körner. Surveys the views of Plato, Aristotle, Leibniz & Kant concerning propositions and theories of applied and pure mathematics. Introduction. Two appendices. Index. 198pp. 5⅜ × 8½. 25048-2 Pa. $6.95

THE DEVELOPMENT OF MODERN CHEMISTRY, Aaron J. Ihde. Authoritative history of chemistry from ancient Greek theory to 20th-century innovation. Covers major chemists and their discoveries. 209 illustrations. 14 tables. Bibliographies. Indices. Appendices. 851pp. 5⅜ × 8½. 64235-6 Pa. $17.95

THE FOUR-COLOR PROBLEM: Assaults and Conquest, Thomas L. Saaty and Paul G. Kainen. Engrossing, comprehensive account of the century-old combinatorial topological problem, its history and solution. Bibliographies. Index. 110 figures. 228pp. 5⅜ × 8½. 65092-8 Pa. $6.00

CATALYSIS IN CHEMISTRY AND ENZYMOLOGY, William P. Jencks. Exceptionally clear coverage of mechanisms for catalysis, forces in aqueous solution, carbonyl- and acyl-group reactions, practical kinetics, more. 864pp. 5⅜ × 8½. 65460-5 Pa. $18.95

PROBABILITY: An Introduction, Samuel Goldberg. Excellent basic text covers set theory, probability theory for finite sample spaces, binomial theorem, much more. 360 problems. Bibliographies. 322pp. 5⅜ × 8½. 65252-1 Pa. $8.95

LIGHTNING, Martin A. Uman. Revised, updated edition of classic work on the physics of lightning. Phenomena, terminology, measurement, photography, spectroscopy, thunder, more. Reviews recent research. Bibliography. Indices. 320pp. 5⅜ × 8¼. 64575-4 Pa. $7.95

PROBABILITY THEORY: A Concise Course, Y.A. Rozanov. Highly readable, self-contained introduction covers combination of events, dependent events, Bernoulli trials, etc. Translation by Richard Silverman. 148pp. 5⅜ × 8¼. 63544-9 Pa. $5.95

THE CEASELESS WIND: An Introduction to the Theory of Atmospheric Motion, John A. Dutton. Acclaimed text integrates disciplines of mathematics and physics for full understanding of dynamics of atmospheric motion. Over 400 problems. Index. 97 illustrations. 640pp. 6 × 9. 65096-0 Pa. $17.95

STATISTICS MANUAL, Edwin L. Crow, et al. Comprehensive, practical collection of classical and modern methods prepared by U.S. Naval Ordnance Test Station. Stress on use. Basics of statistics assumed. 288pp. 5⅜ × 8½. 60599-X Pa. $6.00

WIND WAVES: Their Generation and Propagation on the Ocean Surface, Blair Kinsman. Classic of oceanography offers detailed discussion of stochastic processes and power spectral analysis that revolutionized ocean wave theory. Rigorous, lucid. 676pp. 5⅜ × 8½. 64652-1 Pa. $16.95

STATISTICAL METHOD FROM THE VIEWPOINT OF QUALITY CONTROL, Walter A. Shewhart. Important text explains regulation of variables, uses of statistical control to achieve quality control in industry, agriculture, other areas. 192pp. 5⅜ × 8½. 65232-7 Pa. $6.95

THE INTERPRETATION OF GEOLOGICAL PHASE DIAGRAMS, Ernest G. Ehlers. Clear, concise text emphasizes diagrams of systems under fluid or containing pressure; also coverage of complex binary systems, hydrothermal melting, more. 288pp. 6½ × 9¼. 65389-7 Pa. $10.95

STATISTICAL ADJUSTMENT OF DATA, W. Edwards Deming. Introduction to basic concepts of statistics, curve fitting, least squares solution, conditions without parameter, conditions containing parameters. 26 exercises worked out. 271pp. 5⅜ × 8½. 64685-8 Pa. $7.95

DE RE METALLICA, Georgius Agricola. The famous Hoover translation of greatest treatise on technological chemistry, engineering, geology, mining of early modern times (1556). All 289 original woodcuts. 638pp. 6¾ × 11.
60006-8 Pa. $17.95

SOME THEORY OF SAMPLING, William Edwards Deming. Analysis of the problems, theory and design of sampling techniques for social scientists, industrial managers and others who find statistics increasingly important in their work. 61 tables. 90 figures. xvii + 602pp. 5⅜ × 8½.
64684-X Pa. $15.95

THE VARIOUS AND INGENIOUS MACHINES OF AGOSTINO RAMELLI: A Classic Sixteenth-Century Illustrated Treatise on Technology, Agostino Ramelli. One of the most widely known and copied works on machinery in the 16th century. 194 detailed plates of water pumps, grain mills, cranes, more. 608pp. 9 × 12. (EBE)
25497-6 Clothbd. $34.95

LINEAR PROGRAMMING AND ECONOMIC ANALYSIS, Robert Dorfman, Paul A. Samuelson and Robert M. Solow. First comprehensive treatment of linear programming in standard economic analysis. Game theory, modern welfare economics, Leontief input-output, more. 525pp. 5⅜ × 8½.
65491-5 Pa. $13.95

ELEMENTARY DECISION THEORY, Herman Chernoff and Lincoln E. Moses. Clear introduction to statistics and statistical theory covers data processing, probability and random variables, testing hypotheses, much more. Exercises. 364pp. 5⅜ × 8½.
65218-1 Pa. $8.95

THE COMPLEAT STRATEGYST: Being a Primer on the Theory of Games of Strategy, J.D. Williams. Highly entertaining classic describes, with many illustrated examples, how to select best strategies in conflict situations. Prefaces. Appendices. 268pp. 5⅜ × 8½.
25101-2 Pa. $5.95

MATHEMATICAL METHODS OF OPERATIONS RESEARCH, Thomas L. Saaty. Classic graduate-level text covers historical background, classical methods of forming models, optimization, game theory, probability, queueing theory, much more. Exercises. Bibliography. 448pp. 5⅜ × 8¼.
65703-5 Pa. $12.95

CONSTRUCTIONS AND COMBINATORIAL PROBLEMS IN DESIGN OF EXPERIMENTS, Damaraju Raghavarao. In-depth reference work examines orthogonal Latin squares, incomplete block designs, tactical configuration, partial geometry, much more. Abundant explanations, examples. 416pp. 5⅜ × 8¼.
65685-3 Pa. $10.95

THE ABSOLUTE DIFFERENTIAL CALCULUS (CALCULUS OF TENSORS), Tullio Levi-Civita. Great 20th-century mathematician's classic work on material necessary for mathematical grasp of theory of relativity. 452pp. 5⅜ × 8½.
63401-9 Pa. $9.95

VECTOR AND TENSOR ANALYSIS WITH APPLICATIONS, A.I. Borisenko and I.E. Tarapov. Concise introduction. Worked-out problems, solutions, exercises. 257pp. 5⅜ × 8¼.
63833-2 Pa. $6.95

TENSOR CALCULUS, J.L. Synge and A. Schild. Widely used introductory text covers spaces and tensors, basic operations in Riemannian space, non-Riemannian spaces, etc. 324pp. 5⅜ × 8¼. 63612-7 Pa. $7.95

A CONCISE HISTORY OF MATHEMATICS, Dirk J. Struik. The best brief history of mathematics. Stresses origins and covers every major figure from ancient Near East to 19th century. 41 illustrations. 195pp. 5⅜ × 8½. 60255-9 Pa. $7.95

A SHORT ACCOUNT OF THE HISTORY OF MATHEMATICS, W.W. Rouse Ball. One of clearest, most authoritative surveys from the Egyptians and Phoenicians through 19th-century figures such as Grassman, Galois, Riemann. Fourth edition. 522pp. 5⅜ × 8½. 20630-0 Pa. $9.95

HISTORY OF MATHEMATICS, David E. Smith. Nontechnical survey from ancient Greece and Orient to late 19th century; evolution of arithmetic, geometry, trigonometry, calculating devices, algebra, the calculus. 362 illustrations. 1,355pp. 5⅜ × 8½. 20429-4, 20430-8 Pa., Two-vol. set $21.90

THE GEOMETRY OF RENÉ DESCARTES, René Descartes. The great work founded analytical geometry. Original French text, Descartes' own diagrams, together with definitive Smith-Latham translation. 244pp. 5⅜ × 8½. 60068-8 Pa. $6.95

THE ORIGINS OF THE INFINITESIMAL CALCULUS, Margaret E. Baron. Only fully detailed and documented account of crucial discipline: origins; development by Galileo, Kepler, Cavalieri; contributions of Newton, Leibniz, more. 304pp. 5⅜ × 8½. (Available in U.S. and Canada only) 65371-4 Pa. $8.95

THE HISTORY OF THE CALCULUS AND ITS CONCEPTUAL DEVELOP-MENT, Carl B. Boyer. Origins in antiquity, medieval contributions, work of Newton, Leibniz, rigorous formulation. Treatment is verbal. 346pp. 5⅜ × 8½. 60509-4 Pa. $7.95

THE THIRTEEN BOOKS OF EUCLID'S ELEMENTS, translated with introduction and commentary by Sir Thomas L. Heath. Definitive edition. Textual and linguistic notes, mathematical analysis. 2,500 years of critical commentary. Not abridged. 1,414pp. 5⅜ × 8½. 60088-2, 60089-0, 60090-4 Pa., Three-vol. set $29.85

GAMES AND DECISIONS: Introduction and Critical Survey, R. Duncan Luce and Howard Raiffa. Superb nontechnical introduction to game theory, primarily applied to social sciences. Utility theory, zero-sum games, n-person games, decision-making, much more. Bibliography. 509pp. 5⅜ × 8½. 65943-7 Pa. $11.95

THE HISTORICAL ROOTS OF ELEMENTARY MATHEMATICS, Lucas N.H. Bunt, Phillip S. Jones, and Jack D. Bedient. Fundamental underpinnings of modern arithmetic, algebra, geometry and number systems derived from ancient civilizations. 320pp. 5⅜ × 8½. 25563-8 Pa. $7.95

CALCULUS REFRESHER FOR TECHNICAL PEOPLE, A. Albert Klaf. Covers important aspects of integral and differential calculus via 756 questions. 566 problems, most answered. 431pp. 5⅜ × 8½. 20370-0 Pa. $7.95

CHALLENGING MATHEMATICAL PROBLEMS WITH ELEMENTARY SOLUTIONS, A.M. Yaglom and I.M. Yaglom. Over 170 challenging problems on probability theory, combinatorial analysis, points and lines, topology, convex polygons, many other topics. Solutions. Total of 445pp. 5⅜ × 8½. Two-vol. set.
Vol. I 65536-9 Pa. $6.95
Vol. II 65537-7 Pa. $6.95

FIFTY CHALLENGING PROBLEMS IN PROBABILITY WITH SOLUTIONS, Frederick Mosteller. Remarkable puzzlers, graded in difficulty, illustrate elementary and advanced aspects of probability. Detailed solutions. 88pp. 5⅜ × 8½.
65355-2 Pa. $3.95

EXPERIMENTS IN TOPOLOGY, Stephen Barr. Classic, lively explanation of one of the byways of mathematics. Klein bottles, Moebius strips, projective planes, map coloring, problem of the Koenigsberg bridges, much more, described with clarity and wit. 43 figures. 210pp. 5⅜ × 8½. 25933-1 Pa. $4.95

RELATIVITY IN ILLUSTRATIONS, Jacob T. Schwartz. Clear nontechnical treatment makes relativity more accessible than ever before. Over 60 drawings illustrate concepts more clearly than text alone. Only high school geometry needed. Bibliography. 128pp. 6⅛ × 9¼. 25965-X Pa. $5.95

AN INTRODUCTION TO ORDINARY DIFFERENTIAL EQUATIONS, Earl A. Coddington. A thorough and systematic first course in elementary differential equations for undergraduates in mathematics and science, with many exercises and problems (with answers). Index. 304pp. 5⅜ × 8½. 65942-9 Pa. $7.95

FOURIER SERIES AND ORTHOGONAL FUNCTIONS, Harry F. Davis. An incisive text combining theory and practical example to introduce Fourier series, orthogonal functions and applications of the Fourier method to boundary-value problems. 570 exercises. Answers and notes. 416pp. 5⅜ × 8½. 65973-9 Pa. $8.95

THE THEORY OF BRANCHING PROCESSES, Theodore E. Harris. First systematic, comprehensive treatment of branching (i.e. multiplicative) processes and their applications. Galton-Watson model, Markov branching processes, electron-photon cascade, many other topics. Rigorous proofs. Bibliography. 240pp. 5⅜ × 8½. 65952-6 Pa. $6.95

AN INTRODUCTION TO ALGEBRAIC STRUCTURES, Joseph Landin. Superb self-contained text covers "abstract algebra": sets and numbers, theory of groups, theory of rings, much more. Numerous well-chosen examples, exercises. 247pp. 5⅜ × 8½. 65940-2 Pa. $6.95

Prices subject to change without notice.

Available at your book dealer or write for free Mathematics and Science Catalog to Dept. GI, Dover Publications, Inc., 31 East 2nd St., Mineola, N.Y. 11501. Dover publishes more than 175 books each year on science, elementary and advanced mathematics, biology, music, art, literature, history, social sciences and other areas.